William Parker Foulke

Remarks on Cellular Separation

William Parker Foulke

Remarks on Cellular Separation

ISBN/EAN: 9783337297046

Printed in Europe, USA, Canada, Australia, Japan

Cover: Foto ©berggeist007 / pixelio.de

More available books at **www.hansebooks.com**

REMARKS

ON

CELLULAR SEPARATION.

READ BY APPOINTMENT OF THE AMERICAN ASSOCIATION FOR THE
IMPROVEMENT OF PENAL AND REFORMATORY INSTITUTIONS,
AT THE ANNUAL MEETING IN NEW YORK,
NOVEMBER 29, 1860,

BY

WILLIAM PARKER FOULKE,

OF PHILADELPHIA.

PHILADELPHIA:

1861.

Dear Sir,

We take great pleasure in communicating to you the enclosed resolution, adopted unanimously by the Acting Committee of the Philadelphia Society for Alleviating the Miseries of Public Prisons, on Thursday last.

We feel persuaded that the publication of your valuable essay will greatly promote the cause which our Society has so much at heart,—the extension of the Pennsylvania system of prison discipline.

We remain,
Very truly, yours,
JAMES J. BARCLAY, *Pres't.*
JOHN J. LYTLE, *Secr'y.*

Wm. PARKER FOULKE, Esq.

At a stated meeting of the Acting Committee of the Philadelphia Society for Alleviating the Miseries of Public Prisons, held 12th month 20, 1860, the following resolution was adopted:

Resolved, That William Parker Foulke be requested to furnish a copy of the essay read at the late meeting of the American Association for the Improvement of Penal and Reformatory Institutions, held in New York, and that the same be stereotyped and published under the direction of the committee on the distribution of the Journal; and also that a copy be bound up and distributed with the January number of the Journal.

(Extracted from the minutes.)
JOHN J. LYTLE, *Secretary.*

December 28, 1860.

GENTLEMEN,

The American Association for the Improvement of Penal and Reformatory Institutions, at whose request my essay was prepared, has no permanent fund; and its purpose in relation to the essay was answered by the reading of it at the recent meeting in New York.

Since the executive board of the Philadelphia Society is of opinion that the essay may be of further use, I cheerfully consent to its publication, relying upon its readers for an indulgent consideration of the unfavorable circumstances in which, as they will learn, it was written.

I am,

Very truly, yours,

WM. PARKER FOULKE.

To Messrs. JAMES J. BARCLAY, *Pres't.*

JOHN J. LYTLE, *Sec'y.*

MR. PRESIDENT:

My appointment to deliver to the Association an
address upon the SEPARATE SYSTEM OF PRISON DISCI-
PLINE, imposes an obligation; yet how to fulfil it, is
for me a difficult question. The subject has become
an old one among jurisprudents, and among adminis-
trators of penal law. Since the latter part of the
last century, it has been discussed by a constantly
increasing number of students. In Pennsylvania,
the opinion of the promoters of convict-separation
found some expression in prison construction and
management almost seventy years ago, upon the
recommendation of the Philadelphia Society for
Alleviating the Miseries of Public Prisons; and their
measures had been preceded by changes having a
like object in a few British prisons. During the
next twenty-five years, the subject continued to
occupy the attention of many thoughtful men. The
government of Pennsylvania enlarged the facilities
for the administration of the separate method by
the erection of two large penitentiaries, one, under
the law of 1818, for the Western District, and

1

the other, under the law of 1821, for the Eastern
District; and New York, adopting other conclusions
respecting the means of penal discipline, established
her great experimental prison at Auburn. Scarcely
had these institutions been authorized, when there
sprang up in Massachusetts, in the year 1825, the
Boston Prison Discipline Society, having for one of
its objects the investigation of the penitentiary
question, but avowing its preference for the plan of
convict-congregation. Thus, two associations, at
Philadelphia and at Boston, with experiments assumed to be typical, came into controversy upon the
evidence. The sincerity of each insured the maintenance of discussion. No sooner had the separate
penitentiaries begun to make annual reports, than
the interest of Europe was awakened in a remarkable manner. In 1831 the French Government, in
1832 the British, and in 1834 the Prussian, sent to
the United States commissioners to examine the
most important of our prisons. France repeated her
inquiry among us. Belgium and Russia authorized
a like inspection on their own behalf. From the
commencement of these investigations, the public
discussion of the question between Auburn and
Philadelphia went on abroad with great vigor. In
fact, the question may almost be said to have been
transferred to Europe. The commissioners were men
of eminent fitness for their special duty. Their
reports underwent a searching scrutiny; they were
debated in the executive councils and in the legislatures of their respective nations. A new lite-

rature seemed to be forming. Not only in the bureaux of government, but in the halls of science, prison discipline became a familiar theme. Volume followed volume from the press; so that in the year 1846, when a new era was opened, there had been accumulated a library of works, in the principal languages of Europe, upon this question so peculiar to modern times. The range of argument continually widened, until it embraced penology in its largest sense; and the philosophy of penitentiaries took its due place in jurisprudence. It was in the year last mentioned that there assembled in Germany, at Frankfort-on-the-Maine, a congress of men the most capable in Europe for the profound discussion of penal systems. That was not an assembly of a few persons, citizens of one State. From England, France, Sweden, Belgium, Holland, Denmark, Russia, Prussia, and various other parts of Germany, came those who had been the leaders of reform for their respective nationalities; commissioners to whom had been intrusted the duty of foreign visitation, legislators, councillors of state, judges, inspectors-general, architects, medical officers, chaplains of prisons, members of voluntary associations—such observers and writers as Julius, Mittermaier, Suringar, Ducpetiaux, Moreau-Christophe, David, Russell, Varrentrapp. At Brussels, in the following year, their truly learned and dispassionate discussions were resumed. Meantime there had been formed in New York a third American association, the members of which were pledged to no system, and proposed to themselves a liberal scope. Annual reports from that association were added to our collections. The Boston Society continued its yearly contributions. The Philadelphia Society, since 1845, issued a quarterly journal. Official reports at home and abroad were still multiplied, and every phase of experiment and of opinion found public expression. It is after all this long series, Mr. President, that I am invited to address you, in one hour's discourse, upon the system of convict-separation. Were I able to review all the evidence which has been thus accumulated, to supply new facts which have not been included in previous essays, and to give to the whole a rigorous analysis, this would be to add a new volume to your library—and probably it would not satisfy you.

Not only these considerations, but the lateness of the date—little more than a month ago—at which I received notice of my appointment, assures me that no such task has been intended for me. Permit me to add, Mr. President, not for my own personal advantage, but to acquit me of any seeming defect of diligence, that your notice reached me in the country, that I had very imperfect means of reference, and that such time as I could bestow was snatched from necessary occupations, and was further restricted by the cares incident to the removal of my family to the city for the winter. In fact, more than half of this manuscript has been written during fractions of the last five days.

In the selection of topics, I have reflected that hitherto there has not been in the United States such an opportunity for interchange of explanations as you hope to procure by means of this Association. In the year 1847, the society of New York issued a circular inviting a meeting of officers of prisons and other persons interested in the subject of penal reform. I had the good fortune to attend the sessions of that convention; which was, perhaps, not without some fruits, but which, by reason of the small number of its members, failed to engage the public attention. A similar, and more decisive, fate awaited the convention which was summoned at Philadelphia in the following year. By manifestations at those meetings, as well as by visits to the penitentiaries of different States, and by the printed discussions which had come under my notice, I was convinced that, in many cases, serious difficulties had been occasioned by misapprehension of the nature of the method which has been so long supported by the Philadelphia Society, and of the means by which that method is to be enforced.

The history of the controversy between its friends and those of the congregate method abounds with examples of such misapprehension, even in quarters which one might have supposed to be protected from it by ample access to the proper sources of information.

The very official reports from some of our State penitentiaries bear witness, by their allegations, by their phraseology itself, that the plan of convict-separation is not conceived as we view it; and that the details of evidence are arranged by processes differing from those which we regard as legitimate.

If this Association be in great part composed, as we have all desired that it shall be, of officers of prisons, it appears particularly important to prevent erroneous estimates of the problem which we offer for their solution. Many of them have been for the first time called to the contemplation of penal discipline by the fact of their appointment to places of trust in its administration. Confined by their official duties to the routine of prison government, during an uncertain tenure of their places, they have not had the motive to enter upon general penological inquiries. To many of them there can have been no opportunity for reference to any record of the discussions which have been maintained; and their opinions, even if provisionally held, must have been qualified by the associations in which they have been reared.

Unfortunately, it is not to these alone that the need of fresh explanation is restricted. Every year we find renewed statements which have been in vain contradicted as often as renewed, by which the public mind is misled, and by which a just solution of the leading question is postponed. In formal essays of practised writers, as well as in paragraphs in the daily newspapers, these errors are repeated. Shall I not then best meet the present exigence if I briefly state—

I. WHAT IS PROPERLY MEANT BY THE SEPARATE
SYSTEM;

II. WHAT ARE BELIEVED TO BE ITS PECULIAR AD-
VANTAGES;

III. WHAT ARE THE PRINCIPAL OBJECTIONS MADE TO
IT, and WHAT ANSWERS ITS FRIENDS MAKE TO THESE
OBJECTIONS; and it may be a suitable conclusion to
mention,

IV. SOME OF THE OBJECTS FOR WHICH, DURING THE
PENDENCY OF THE QUESTION OF ITS MERITS, ITS FRIENDS
MAY JOIN WITH ITS OPPONENTS IN BEHALF OF A
GENERAL PENAL REFORM.

I. What is meant by the SEPARATE SYSTEM?
Shall I begin by noticing some things which that
system is not? It is not the "solitary" system—
unbroken isolation without labor; nor is it solitude
with labor. It is not Olmutz nor the Bastille, as
General La Fayette was led to believe; nor is it, as
the benevolent Mr. Roscoe was at one time persuaded,
"a new invention, heard at first with horror, but
gradually revealed to the public, till at length it has
been unblushingly brought forward and recommended
to the adoption of states and communities as an
advisable and even philanthropic measure."

It is not any one of those "experiments" tried in
New York, at Auburn, in Maine, in Virginia, in
New Jersey; experiments which remained during
many years prominent sources of argument and
even vilification against the friends of separation,

and the influence of which is still evident in quarters where most of the facts have been forgotten.

It is not typified by that class of experiments, of which we have many examples, to which belongs the history of a warden of the Massachusetts State Prison, who, not many years ago, selected a dangerous fellow, shut him up in a large cell, which was furnished with a bed "and every thing to make him comfortable," and there employed him in shoemaking. At the end of several weeks " his humility and constant complaints of loneliness and misery" so worked upon the officer's feelings that the determination of this benevolent man was conquered, and he was obliged to restore the convict to the yard and workshop! Of course, the history of the thousands who had been subjected to the separate discipline in Pennsylvania went for nothing. Had not the warden seen with his own eyes, heard with his own ears? It is worth your inquiring how many of such TESTS are annually applied, and what influence they exercise upon the opinions of officers.

Again, the separate system is not, and cannot ever be, represented by such isolations as are mentioned in the report of the Massachusetts State Prison for the year ending September, 1858; where we are told that a greater or less number of convicts who are considered dangerous are always in close confinement without labor. Even with labor and instruction and visitation, such exceptional applications of our rule would be mischievous. This has been proved in Switzerland. in France, and

even at Pentonville in England, as well as elsewhere.

Again, the separate system is not that which has
been maintained during a long time in the large
penitentiary of New Jersey, where gradually the
distinguishing characteristics of the separate method
fell into such neglect that it became not easy to say
what remained, except some apparently useless and
troublesome formalities, which have been, naturally,
fast yielding to the supposed economical advantages
of common workshops and the contract-system of
labor.

I must add that the separate system is not that
maintained at the Moyamensing prison at Philadelphia,—a county jail constructed in defiance of the
requisites of our discipline, and subsequently in part
converted into a State penitentiary for convicts, as
well as continued as a jail for persons under arrest
merely. It is now an establishment where vagrants,
drunkards, and persons detained for trial, as well as
convicts, are confined in one building. As many
as twenty thousand commitments have been made
in a single year. An excellent board of inspectors
have contrived to maintain a certain degree of advantage, notwithstanding the physical difficulties
with which they have to contend; but the institution offers no proper exemplification of our idea of
convict-separation.

Let me interpose another preliminary caution.
The word SYSTEM has received various acceptations
in connection with the reform of prisons. It has

been made to comprehend both a code of penal law and its administration in all details; it has been applied in a more limited sense to the general plan of prisons and their internal management; it has been used with reference to the plan alone; and it is in the last mentioned case that the restriction of it to the characteristic feature of a plan has originated. Now, without expecting to correct any of these various and inconsistent employments of the word SYSTEM, it may at least be required for our clear understanding of one another that the particular sense which we give to the word shall be defined and maintained on each occasion of its use. This will prevent confusion of the essentials of a plan with any one form of its administration. For example, if I were to speak of the congregate system, and were to cite constantly the penitentiary at Sing Sing as its type, our friends from Charlestown, Massachusetts, would doubtless very soon remind us that fair reasoning requires a more abstract method. We should be told that the proper subject of consideration is that which distinguishes the congregate discipline in general; and that the mode of administration is an accident, which may or may not have a permanent value in discussion.

You may think these cautions superfluous. The history of the controversy does not so teach us. I think it no disrespect to any one to say that some of the forms of misapprehension already adverted to, may exist even here in the midst of us. In

Europe, certainly, with all the advantages of
public instruction which have been available to
educated persons; in the face of legislative debates,
reports of commissioners, ministerial circulars, and
other aids to precise knowledge, not only the
evidence, but the very conditions of the question
of discipline are widely misunderstood. I shall
therefore even go further to remove possible impediments from our way.

What is it, then, that you conceive as embraced
by the term SEPARATE SYSTEM when you bring convict-separation into competition with the congregate
method? Is it quite clear that the image of one
selected prison is not always dominant on each side
as the thing to be tested? When you compare the
financial economy at one place with that at another,
have you secured every due allowance for elements
of proper cost in construction, in food, in instruction? If you array one bill of mortality or disease against another, have you carefully estimated
the discrepancies of population, age, color, and
degree of minute inspection; also the length of
sentences, both absolutely, and in their special
relations to the peculiar discipline? Have you, in
short, put your prisons upon an equality in other
respects—in the respects common to both methods
—before you begin to estimate the evidence proper
to the feature which distinguishes them, that which
is alone the subject of dispute? A strong illustration is at hand. A gentleman holding one of the
highest official positions in one of the United States

has said to me, "Sir, the penitentiary of my State is a disgrace to a civilized community. The labor of the prisoners is let out to contractors, who pay so many thousands of dollars to the public treasury; there are officers enough to keep the prisoners from running away—and that is all we know about it. The State makes some money, and this keeps everybody quiet." Would any of you who have visited the Eastern Penitentiary, think of settling the question of financial economy by a comparison of its earnings with those of the other prison just described?

Let us suppose that we have accomplished all the conditions which naturally precede the development of our question; that the site and structure of the buildings are what each method requires; that we have provided a due supply of suitable food and sufficient intellectual and moral teaching; that our labor is in just relation not only to the cost of maintenance, but also to the exigencies of rational and humane discipline; that our sentences are duly proportioned; that our officers are selected and paid with a liberality which will secure, for the most part, the requisite grade of character; that the county jails, the places for detention before trial, are in harmony with our penitentiaries for convicts. Thus far we have common objects, and there has arisen properly no question between us.

At this stage we of Pennsylvania say that, in the midst of all these provisions, you are thwarting your own purposes; that by associating your convicts together you promote vice and crime, and hinder the salutary operations of penal discipline; and that you inflict evil consequences not designed by the law, and to which your prisoners ought not to be forcibly and authoritatively exposed.

Thus, then, arises our question, which is simply whether or not it is necessary, or, what is here the same thing, proper, to compel prisoners to be associated WITH ONE ANOTHER.

Perhaps it will be objected that the mode of statement should be reversed; that the inquiry is whether or not men should be forcibly separated from companionship, isolated, so to speak; that the law of our nature impels us to society, and that, consequently, the necessity for separating convicts from one another should be proved by the friends of cellular imprisonment.

Mr. President, I have deliberately chosen my form of statement, first, in order to exhibit a monstrous fallacy, which has vitiated from the beginning of the controversy the reasoning of most of the friends of congregation; secondly, in order to show to you that the burthen of exceptional proof in fact rests upon them, and not upon the supporters of convict-separation.

The fallacy consists in a confusion of the proposition that "men require SOCIETY," with the other proposition that "a SOCIETY OF CONVICTS is necessary." Concede that man is a "social being," how does the necessity of his association with CONVICTS follow? Are we to infer the expediency of a society

2

of rogues, from that of a society of men in general?
to make the social nature of the race a proof that
those individuals who have unfitted themselves for
any society, shall, for their improvement, form a
community by themselves in prison?

The burthen of proof results from the state of
the case, from those facts which are conceded on all
sides; viz., that the moral and intellectual condition
of convicts is exceptional; that their vicious education and habits demand a treatment different
from that which we give to the virtuous; that their
mode of life has engendered purposes, and wants,
and sympathies, which must be broken up and de-
feated; that the confederacy amongst them must be
dissolved; and that they must be accustomed to the
companionship, the sympathies, the habits, and the
pursuits, of honest people. Thus the most natural
inquiry is, not, why will you separate those men? but
rather, why, having in view a moral reformation, will
you compel them into association with one another?
The very social law which you invoke teaches this
statement; all your other methods of repression
and of education are in harmony with this.

Nay more; the founders of the congregate system
as it has been established in the United States
announced at the outset that their object was the
separation of convicts. Look back upon the reports
of legislative committees, upon the long series of
argumentative reports issued by the Prison Society
of Boston, upon the controversial pamphlets published under the sanction of the Philadelphia

Society, upon all the literature which issued from the press during the period of the establishment of the New York and Philadelphia penitentiaries, and you will see that the parties were all aiming at the separation of prisoners. It was a great reliance of the advocates of the Auburn prison that it was, as they thought, absolutely effective as a means of separation. This will appear natural when it is remembered that from the days of Howard the mischiefs of association constituted the most urgent of the motives to reform in the internal régime of prisons.

It is true that we no longer hear of the possibility of preventing all intercommunication in the common workshops. The dogmatic assertions of former times, supported by the scourge and by an attempt at unwavering severity and supervision, have happily ceased; and amongst the intelligent keepers of the best congregate penitentiaries we no longer find champions of the "rule of silence" as it was once vaunted. Nevertheless, it remains also true that the discipline of those penitentiaries does not contemplate the association of prisoners as a means of reform. It enforces a penalty of confinement and labor; but it professedly submits to the aggregation of its subjects only because this is supposed to be the means of superior health and economy. I have not, then, misstated our question; and by this brief review it will be apparent that even with respect to the motive and primary effort of both parties there is still identity. We are all aiming to bring about such an interception of evil communications as shall leave men who have been convicted of crime open to our reformatory influences, without exposure to the mischiefs of depraving companionship.

The method adopted in Pennsylvania goes to the root of this problem, and it does not trust the desired separation either to the caution or the good resolution of the convicts on the one hand, nor to an impossible vigilance of officers upon the other. IT DEMANDS THAT THE PRISONERS SHALL NOT BE ASSOCIATED, BY DAY OR BY NIGHT, WITH ONE ANOTHER. This said, you have all that properly distinguishes that method from any other which contemplates prison discipline as understood in our day.

Persons whose conceptions have been guided only by the antagonistic relations of the two leading modes of imprisonment; persons who think of the long controversy which has been waged, and the active partisan resistance which has been manifested on both sides, may be disposed to regard this definition as too restricted. It may be thought even by some members of this Association that the friends of convict-separation, becoming convinced of the extreme nature of their original plan, have desired to bring it into a more favorable condition, by accommodating its details as far as possible to those of the congregate system—in short, that we have practically conceded the alleged extravagance of our original views of prison life. If this thought were true, at least it must carry us back a great many years to find the supposed modification. For my own part, it is certain that no representation has ever been authoritatively made to me other than that which has been given to you in the foregoing remarks. Fourteen years ago, having been honored by an invitation to attend an anniversary meeting of the New York Prison Association, I had no difficulty in writing as follows:—" We have one inflexible purpose, that of preventing any society of criminals. Beyond this, to whatever arrangement can be made for securing the health of prisoners, or their mental and moral improvement, we set no limit which falls short of the grandest, most christian view of duty from man to man. That the parsimony of governments, and the ignorance or indifference of private persons, will impede the entire fulfilment of that duty, may be reasonably expected; but it would be a sad reproach to the citizens of this republic if no other means could be devised for preserving the mental health of an offender in confinement, than is afforded by his association with other criminals."

At the same time the Philadelphia Prison Society made an official communication, which is signed by its president, vice-presidents, and secretaries, all of whom had long previous training in the traditionary opinions of that society, and in the visitation of the prisons of Philadelphia. They then said, " The separation of prisoners from contaminating influences, and carefully training them by means of judicious instruction, form a portion of the disci-

pline of every prison where reformation is regarded.
The Philadelphia Society for Alleviating the Miseries
of Public Prisons, more than fifty years ago, were
convinced that as evil association corrupts good
morals, so such associations would be deleterious in
an increased measure within the walls of a prison.
They inferred that a career in sin might be retarded, and in many cases terminated, *by removing
an offender from the society of the wicked, and associating him exclusively with the intellectual and virtuous.*
They have never devised, much less attempted, the
separation of prisoners from all society; nor has
such a plan ever been sanctioned at any time by
the Legislature of Pennsylvania."

At the very opening of the Eastern Penitentiary,
more than thirty years ago, the same views were
expressed by one of the officers who signed the
communication just cited, Mr. George W. Smith, in
his "Defence of the Separate System," republished
in 1833 by order of the Philadelphia Society. In
that valuable witness of the opinions and designs
which were entertained and published at the period
just mentioned, the period of most excited partisanship, the author said, " Religious and other instruction will be constantly and regularly administered;
the visits of the virtuous and benevolent permitted
and encouraged, under proper restrictions; unremit-
ted solitude, or separation from *all* society, will not
be, therefore, permitted." Again, "It was never intended by the friends of our system, even by those
who were opposed to the introduction of labor, to

deprive the convicts of exercise, of books of instruction, and of suitable society."

There ought never to have been any misunderstanding on this head. Although the Philadelphia Society has always labored quietly and unobtrusively, paying due respect to the legally constituted officers in every department of administration, and never pushing itself in the way of ordinary functions for which others were responsible to the public, yet from the beginning of penal reform in Pennsylvania, down to the present time, the footprints of that society may be traced in advance of every important change in our penal system. In the modification of penalties, in the establishment of prisons, in the regulation of their construction, at every stage, you find a memorial and a committee of that society. The revision of our entire criminal code, which has been made within the last two years by a commission authorized by the legislature, was undertaken at that society's instance. For a serious inquirer, therefore, it is easy to ascertain what views of discipline have always prevailed in Pennsylvania.

I insist the more upon these preliminary suggestions, because it is in vain to enter upon a discussion unless the subject of it is clearly known. We cannot be tried by the notions, not only vague but variant, which pervade the community; and it is not superfluous to add that the ideas which have been so consistently maintained in Pennsylvania are the same which have been received in Europe

as the proper characteristics of the separate system.
From the visit of the first foreign commissioner, to
the last published foreign discussion, the friends of
convict-separation abroad as well as at home have
avowed the same fundamental conceptions.

It is one of the evils of prejudice, and of a partisanship without due information, that they not
only originate but continually propagate false
notions of the questions which they undertake to
solve. Hence it happens that even the precautions
already briefly noticed are not sufficient to clear
the field for pertinent controversy. It is not only
by a consideration of the meaning of convict-separation by itself, as expressed by its friends in a
positive sense, that we are to learn its value for
discussion. It is one of several proposed means of
discipline, one of which is necessary. The question
is not between one example and an abstract perfect model; but BETWEEN THE SEPARATION AND THE
CONGREGATION OF CONVICTS UNDER THE PRACTICAL
RESTRICTIONS INEVITABLE FOR EACH. Hence, to conceive our definition correctly for practical uses, we
must render our meaning more precise by a reference to those things for which we offer it as a sub-
stitute, and in relation to which it has been chosen.
It is with the certain alternative of a community of
criminals in prison, that you are to accept or reject
the policy of separating those criminals from one
another. What community of criminals? Have you
decided what it is that you would compare with the
principle which we offer to you? Not a discipline

imagined by assembling the particulars of good and rejecting the particulars of evil from all the known prisons—not by confounding or wholly omitting all circumstances which qualify localities, and only the resultant of which, after all requisite explanations, is available to you—not the fancied discipline which won the faith of the early friends of the Auburn plan in the United States—a discipline to which was attributed the power of absolutely isolating the members of a convict society even in the communities into which they were forced. This is exploded everywhere, not only as an advantage, but even as a possibility. Is it, then, a community of criminals amongst whom some communication is to be expected? Is it one in which labor is an instrument of public discipline, held and controlled and applied solely by the officers of the prison; or is it one in which the convicts are let out by the day upon contracts which introduce new agents into the field— agents not appointed for the functions proper to discipline, nor responsible for them in any way beyond a compliance with the rules having reference to safe custody? Do you contemplate the lash as a means of restraint and of correction? All of these topics have a vital connection with the moral and prudential relations of your inquiry, and you must choose amongst them before you can reduce your problem on both sides to the simplicity and completeness which characterize the side already placed before you. I make no comment at this stage upon schemes of classification—schemes through which the idea of separation had its germination and growth—schemes everywhere tried and everywhere distrusted in Europe—schemes which, nevertheless, with a disregard of history not uncommon in this country, are again brought forward. Even as to these we should be compelled to require fresh discrimination; for it is manifest that they must in many cases clash with the contract-system of labor; and they have, in fact, recently been rejected on that account by the Inspectors of the Massachusetts Penitentiary.

II. With these bare hints at sources of the confusion which besets us at the threshold of our investigations, I proceed to a notice of THE ADVANTAGES OF CONVICT-SEPARATION; and, as an indispensable preliminary, I invite your attention to a topic which may be regarded as a touchstone of our preparation. It will not be disputed that a prison, whatever its construction or management, is not one of the ultimate objects of society. It is not for their own sake that penitentiaries are established. They are instruments—means to an end. Our definition, thus far, has been effectual only toward the ascertainment of the nature of the means—the identification of the instrument. It is too plain for argument that the true value of the instrument is to be determined by the relation which it bears to the end for which it is chosen.

If our object is to punish, we shall look for

severity. If our object is to deter by causes of fear, we shall look for agencies which excite public terror. If we hope for pecuniary gain, we shall favor every arrangement which promises to yield to us the largest product of the given human machinery for the given term of use. If we desire the intellectual and moral improvement of our prisoners, we shall select those agencies which are fitted for their education in knowledge and virtue. If we desire ALL of these things, then must we so proportion our adjustment of them as that neither shall interfere with the rightful claims of the others upon our efforts.

I have used the word RIGHTFUL, because, be it remembered, there is still another check upon our judgments. Not only must we determine the end with reference to which we construct prisons and regulate their discipline, but that end must be a justifiable one; our choice of it is in subordination to paramount general laws. We can conceive of a despotism so absolute as to be determined only by its own will in its choice whether of ends or means. We are restricted as to both. There will doubtless be no dissent when I say that it would not be right for our governments to adopt any one of the particular objects just mentioned, as the sole end of its discipline; nor for keepers of prisons to aim at the maintenance of the "rule of silence" by extreme penalties, as was done at Sing Sing; or at a large pecuniary return from prison labor to the neglect of higher considerations, as has been done in many places.

It is necessary, therefore, that at the outset we satisfy ourselves, not only with respect to the nature of the discipline in question, but also with respect to the designs which it ought to subserve.

It is possible that some persons, accustomed only to the execution of a predetermined series of prison rules, persons who have confronted only the immediate practical duties resulting from such rules, may be disposed to regard this, as well as others of our preliminary considerations, as merely speculative. Let us not debate their impressions. If those considerations are to be kept outside of the range of your thoughts, as not constituting a portion of knowledge useful to the administration of prisons, then at least abstain from expressing any opinion upon them; keep them from the official reports of your prison officers; leave them to be discussed by men who make of them a special study, in the light of history and of social philosophy; confine the exhibits of your penitentiaries strictly to the details of management, and avoid all phraseology which may mislead others. If, on the other hand, any one will persist in giving utterance to conclusions which, notwithstanding their profound difficulty and the wide comprehensiveness of their consequences, he has ventured to base upon the narrow circle of experience in a single prison, at least let us require that he come within the range to which his conclusions belong, and that he submit them to the appropriate tests. Dogmatism will not suffice. We must be prepared to reconcile to rightful ends of government whatever discipline we may ultimately prefer. Without such preparation, we can have no claim upon the attention of the legislative bodies whose high authority we invoke to sanction our resolutions by their enactments, no claim upon the confidence of that class of jurisprudents whose wisdom in every well-regulated State inspires the public councils.

In spite, however, of all pretensions to freedom from these speculations, everywhere we find that there are tacitly assumed principles of legislation and of administration which are very questionable—special ends of discipline, which are not in harmony with the conclusions of the best judges, under whatsoever system; and there is maintained an independence of management which is fatal to the idea of a community of counsel amongst experts. These tacit assumptions govern all the reasoning of those who adopt them; and, being covertly suggested, rather than openly presented, to the community, they influence public opinion to an extent that would be impossible were there a formal, precise discussion of their merits.

It will, of course, not be expected of me, on the present occasion, to enter upon the investigation of the nature, the objects, and the just limitations of punishment to be inflicted by civil government; but there are some truths which are recognized by the best thinkers as proper to every plan of penal discipline; and to these I shall confine the exposition which remains to be made.

With respect to the advantages of convict-separation, there are to be distinguished two classes of contestants; the first of whom deny the importance of such separation as is proposed both by the supporters of the Auburn plan and by those of the Pennsylvanian. They assert that, under any régime which will maintain general order in the workshops without extraordinary risks to safe custody, there is no probability of such corrupting communications as would demand a further separation of the prisoners. To this class belonged a former warden of the Massachusetts State Prison, who officially stated that he believed "that the few words which a convict can steal the opportunity to say are full as likely to be good and encouraging as evil and debasing;" and the supervision was at that time less stringent than it now is. To the same class belonged a former warden of the New Jersey State Prison, who allowed to remain open the holes which the convicts worked between the cells, and justified himself by asserting that such communication as would take place through such apertures could do no harm. Opinions like these it would be idle to discuss at this day and before this Association.

The other class consists of the major part of the intelligent advocates of the congregation of prisoners by day. According to them, there should be total separation by night, as we too insist; they further require, however, that prisoners should perform their labor in common workshops, but under a rigid prohibition of intercommunication. To prevent this, rules to the breach of which penalties are affixed, are prescribed to the prisoners. The maintenance of these rules is intrusted in each shop to one or more officers, who have the sole custody of the inmates of it, and who are armed, and are known to the prisoners so to be, for the enforcement of obedience.

Under the separate system, each convict occupies an apartment from which other prisoners are excluded, during both day and night; and custody is secured by the walls and door of that apartment.

Before we reach debatable ground, let us observe some consequences which inevitably follow from the mere difference of the physical conditions which have just been stated; consequences the reality of which must be apparent, whatever we may think of their extent or value, and however they may seem to be overbalanced by other considerations, connected with health or with economy.

In the first place, then, assuming one of our objects to be to hinder intercommunication, the separate system adopts the most efficacious known means. It is to be noticed that this proposition has two aspects. There is an intercourse to be considered which takes place by consent of all parties to it, which is purely voluntary with all; and this would be sought to a certain extent in every prison by a portion of its inmates. For some purposes of intercommunication it would be almost impossible to prevent occasional intelligence between the occupants of adjoining cells. A system of signals by means of blows upon a partition-wall could be established by rogues out of prison, to be practised in case of incarceration. This plan, however, requires that the parties to it shall occupy the same prison and adjoining cells in it; and it must constantly expose them to detection; as a series of knocks sufficiently loud and peculiar to serve the uses of conversation, could not fail to attract the ear of a keeper on one day or other. Besides, the opening of conversation in that way is optional. Every man is free to remain silent and unknown. Some persons have thought that any breaking of the abstract symmetry of the method of prevention is sufficient to reduce the separate prisons to an equality with the congregate ones; and it has been a very frequent answer to the allegations of advantage in this respect, to say that communication takes place in separate prisons; as though this general statement covered all features of security and adaptation, and all measures of success. It would be quite as fair to class all congregate prisons together, and to predicate of them equal appropriateness for discipline because in all of them there is intercommunication in the shops. Here we find another of the fallacies which have served to delay a clear appreciation of the terms of our question. Looking to the mere physical arrangements proposed, it may safely be left to a sober judgment to decide which of them offers the greatest material guards against intercourse.

The separate system, moreover, offers fewer temptations, as well as fewer facilities. It does not bring men together and compel them to sit together, work together, walk together, month after month and year after year, and expect them to be silent. It does not aggregate in close proximity men of like habits, tastes, and sympathies, and then press them to labor under a prohibition of words and looks of recognition and companionship. It places a wall between them, and endeavors to protect each from a knowledge of his neighbors. It aims to concentrate the thoughts of each upon his situation, partly by removing unfavorable external stimulants of old associations, thus sheltering him against all provocations of false pride, and against the evil influences of that co-education in crime by which every natural feeling is interlocked with mischief and is subjected to the domination of every member of a wicked fraternity. In short, it follows the old maxims of education, by removing the accustomed external motives to vicious thought, while it introduces the manners, the ideas, the purposes, and the sympathies which are proper to honest, well-meaning men. There can be no risk in asserting that it employs for this end the most efficacious known means.

Not only does cellular separation guard prisoners against the interferences occasioned by their old associations; it also protects them against provoca-

3

tions which in other modes of confinement grow out
of their relations with their keepers. The force
which restrains is that of the law : the walls and
door of each cell, and not an armed officer, keep the
peace. There is nothing to suggest that idea of
personal antagonism which is developed by the
arrangements of the congregate rooms, and the
fruits of which are shown in personal encounters.
With us the officer is a visitor who supplies food
and the means of employment, rather than a guard
upon whose vigilance and courage and strength
depends the subordination of the men committed to
his oversight. Consequently we have no mutinies,
no riots, no angry assaults upon keepers; and consequently we avoid inciting the minds of our prisoners to schemes of outbreak or revenge—schemes
which occupy many of the heads and hearts in the
congregate shops, and the very thought of which,
promoted by bad companionship, feeds the passions
even of men who would not dare to strike a blow
for their realization.

Again, we avoid placing our prisoners in such
circumstances as deprive the officers of an option in
the application of special punishments. It is well
known that many of the most earnest strugglers for
escape from the ways of crime are men who, from
peculiar sensitiveness or irritability of nature, are the
most ready to yield to momentary impulses. The
hardened villain submits to his fortune, endeavors
to make his place easy, avoids conflicts with those
who control his comforts, and gives comparatively

little trouble. It has become a trite saying in European as well as American penitentiaries, that the worst criminals make the best prisoners. It is obvious that, in the face of a large company of convicts, every violation of discipline, even angry or sulky acts or words, must be in some way noticed. The keeper cannot temporize, nor frame excuses for the offenders; whereas in the separate cell all the modifications which prudence and humane consideration may sanction are practicable in any case, because custody and discipline are not thereby put in jeopardy in other cases. This is a very important consideration.

It is too obvious to require comment, that in the congregate rooms are occasioned some of the incitements which lead to breaches of disciplinary rules; the convicts are stimulated to intercourse, and to devise modes of escaping the vigilance of the officers.

You will observe, too, that as the professed object of all parties to this controversy about systems is, both to remove impediments to moral improvement, and to encourage spontaneous efforts at self-reformation, we accomplish the latter purpose in a large measure by the success which our plan secures for the former. Whatever there is of susceptibility to good recollections, to those of early years, of home, of kindred, of the lessons of any former period, we do not benumb by hourly companionship with depraved men, who revive only evil thoughts, the reminiscences of a vicious life, and of the incitements and the enjoyments of its unbridled license. Every man may be

regarded as carrying in his bosom a duplex character. In every human mind are the elements of a
moral, and those of an immoral, system of ideas.
What determines the preponderance of either in free
society? What encourages, augments, strengthens
either? The recurrence of correspondent motives;
the accumulation and repetition of conformable ideas
and emotions; the presentment of stimuli of like
kinds. If you desire anywhere—at your domestic
hearth—in the community at large—in your solitary chamber—to free yourselves from any habitual
current of thoughts or feelings, what is the course
pursued by you? What is meant when human
beings appeal to the common Father to be not led
into temptation? It is not that they may be hourly
tried by being hourly exposed to sources of the very
temptation which they would escape; but rather
that, in mercy to their feebleness, they may be withdrawn from these, lest old habits and the willing-
ness of an indulged nature may prove too much for
their good resolutions. It would be a mockery of
such a prayer to persist in the companionship which
keeps alive only evil reminiscences and evil wishes
and purposes. Therefore the friends of convict-
separation seek the removal of these, in order that
the germs of self-reform may grow without hin-
drance, and that whatever there is of capability in
this respect shall at least find no impediment in the
administration of the prison itself.

Curiously enough, this very exemption from temptation has been used as an argument against the

separation of criminals from one another. It has been said, both here and in Europe, that inasmuch as prisoners are to re-enter free society after the expiration of their terms of sentence, they ought to be seasonably habituated to resist evil incitements; and that if we prevent their exposure to the risk of further contamination, and allow them to associate only with honest people in prison, we deprive them of a proper opportunity to fit themselves betimes for their career by suitable preparatory training. Foreign as well as American writers have fully answered this extraordinary argument. A constitution shattered by an unhealthy climate would as reasonably be confined to treatment under that climate—a lad depraved by dissolute comrades would as reasonably be forced to maintain hourly association with them in order to accustom him to resist their allurements—as prisoners could be authoritatively kept in contact with one another to prepare them for the habits of an honest life. Of all the claims in behalf of the congregate system, this one is the least in accordance with the common judgments of mankind upon the relation of means to ends in a corrective discipline. As with many physical diseases, so, as has been forcibly remarked, "in the perturbations and disorders of the intellectual and moral nature, a long period must be passed before we can hope to eradicate the evil habits of which these disorders are evidences and effects. There is one indispensable preliminary condition for the treatment of both these classes of morbid states,

—viz.: a removal of the primary and sustaining causes, as far as these depend on external and appreciable circumstances. We should not think much of the professional skill of a physician who contents himself with prescribing medicines to a person suffering from lead-poisoning, but who fails to recommend his patient to withdraw from the manufactory in which he is in continual contact with the poison. Is there more wisdom in those who pretend to reform a criminal while allowing him to associate in any fashion with other criminals, and to imbibe continually the moral poison with which he is already grievously affected?"

It has been said that cellular confinement is not favorable to intellectual and moral instruction. Time has overthrown this objection; but, in the present connection, it is worthy of notice that advantages result necessarily in this respect from the seclusion of convicts from one another. The instructor not only has a choice of time, which is impossible under any other arrangement, and which is of great importance to the individualization of his efforts; he has also in his favor the fact, that his teaching is rendered more acceptable by its coming in a better manner, as an alleviation of punishment.

It is true that it would be *possible*, in a congregate prison, to take the men successively from the shops into a well-lighted apartment of comfortable dimensions, and to allow each of them to receive the recreation or the moral benefit of a lesson; but before this possibility can be converted into a regular practice, there must be alterations both of construction and of labor-contracts, which will change the face of our financial comparisons, as there will be opportunity to notice hereafter.

In a separate cell the intervals between the hours of instruction may be more freely, and probably will be more customarily, given to the reconsideration of what has been learned, than in an associate room where there are many and constant external distractions. With respect to moral reflections, it is well known how easily these are broken up by light words and looks. For many persons, the moving effect of a sermon lasts only as far as the outer door of the church—having decreased at every step toward it, among the recognitions of acquaintances. A grimace from a vicious comrade, or even the sight of his face, may defeat, in the mind of a convict, the influence of an hour of judicious exhortation from a chaplain. It ought not to be forgotten that to every mind there come special seasons of susceptibility and of impulse, when the moral nature is at the flood, and when, if not hindered by external restraints, it seeks the fullest expression away from observing eyes. It is in such epochs of sentiment that are formed, especially by persons who have not been systematically disciplined, those fresh resolutions which, if protected against intrusion and encouraged by appropriate suasives, lead ultimately to genuine repentance and amendment of life. They are fa-

miliar phenomena of our emotional nature. It
needs no repetition of instances to show how little
opportunity can be afforded to them in the congregate rooms.

That employment is more welcome, and labor
more spontaneous, where they are the means of
relief to the loneliness of a cell, than where ex-
ertion is coerced during a fixed number of hours in
the midst of an assembly of convicts and with a
never-ceasing exhibition of force in the back-ground,
needs not to be proved here. If we are to reconcile
to steady occupation, men most of whom owe their
incarceration to a dislike of it, we must accustom
them to find in labor a comfort which they have
not known; to obtain voluntarily from it, by habitual application, protection from evil thoughts
and from the natural results of idleness, for the
first time clearly manifested to them.

There remains to be mentioned an advantage the
value of which has been generally underrated by
the supporters of the congregate method, and perhaps sometimes overrated by those of the plan of
separation: it is the guaranty given to every inmate of a separate prison that he shall not be
exposed to the formation of new acquaintanceships
among criminals. If this topic could be freed from
connection with the general controversy upon penitentiary systems, and if the question could be
nakedly put whether any scheme of discipline of
offenders against the laws should require absolutely
that its subjects should be compelled to widen the

circle of their associations so that this should include every known class of criminals, there could be little difference of opinion. It would not be contended that any person, whatever the grade of his offence, should, under pretence of his amendment, be brought into larger contact with the dangerous classes. We should rather both counsel and assist him to cut short his communications with his old associates, and to avoid placing himself within reach of men who, upon his liberation, would put in peril his good resolves and multiply for him the sources of temptation. Knowing, too, how every reformed prisoner is exposed to the threats as well as the arts of persistent criminals, we should endeavor to shield each inmate of our penitentiaries against such knowledge of his person as would place him at the mercy of unprincipled men. The congregate system leaves no option to convicts. Whatever the previous character or social relations, whatever the grade of offence, whatever the age, and whether the confinement be upon the first or the tenth conviction, every man is thrust indiscriminately into the company of the common workshops, is generalized by one standard, and is forced to feel the equality of rogues.

You may laugh at the idea of a convict being too nice for association with other convicts; but you will not thereby destroy the fact which has been many times attested from congregate penitentiaries, that the compulsory exposure of prisoners in the workshops, and their daily contact with criminals of all grades, inflict a moral wound which is often incurable by all the remedial influences of your discipline. I have myself heard such attestations, in such penitentiaries, from the lips of men whose self-respect had been cruelly invaded in this manner.

If there were no other reason to plead for the separation of convicted persons, I would hold, Mr. President, to the sufficiency of this. I would protest, in the name of humanity, in the interest of the righteousness of our laws, against this abuse of the power of the State.

We say that in this respect there is an exemption offered, rather the performance of a most solemn duty secured, by cellular separation. In this respect, at least, the majesty of public justice is not mocked by a process which confounds all ideas of consistent discipline, and which to the formal penalty prescribed by our code adds a wrong which no proper end of government can sanction. In vain have methods of classification been tried with the hope of overcoming this mischief. One by one, wherever tried, they have been abandoned, as deceptive and as tending at once to produce on the part of prisoners hypocrisy and a concentration of evil, and on the part of officers delusion and danger.

To persons accustomed to regard convicts simply as a CRIMINAL CLASS, who are to be controlled by force and punished according to the literal tenor of judicial sentence, the observations just made, and indeed all others looking to the individual treatment of prisoners, have ordinarily appeared to be over-refinements of an unpractical, sentimental, theoretical philanthropy. Yet is it not clear that if each convict is to be punished and reformed, or deterred, it is AS AN INDIVIDUAL that he is to undergo our discipline? You cannot educate men in masses: though they may be to the eye an aggregation, their susceptibility, their destiny, are single. Adopt what method you please, it can affect each man only with reference to his own peculiar qualities and condition and prospects. Every prisoner carries with him his own little world of associations. His past life, his future life, are his own. Any successful method of moral reproof or education must be conformed to his consciousness; and, as far as reformation is voluntary, it is to grow out of the actual state of the man. Whether you educate men in prisons, or children under your own roof, the reason is the same; the same laws govern the procedure. There is not time to elaborate this topic, nor to adduce all the well-settled principles upon which its proper treatment depends. Whatever estimate you may make of its value, the fact is incontestable that cellular separation protects discipline against serious interferences which result from compulsory aggregation.

I have frequently been met by the suggestion that public degradation to a common rank with convicts is one of the most efficacious of punishments, and one of the most operative means for deterring from crime; and that the susceptibility of prisoners to such degradation gives to the congregate discipline a great advantage. This might be true, to some extent, if the sole object of penitentiary discipline were to punish; which it confessedly is not. With respect to prevention, it is obvious that the alleged advantage must exist, if at all, chiefly in those cases in which, from the education, character, and previous life of the offender, his susceptibility really receives its chief shock from the idea of public conviction and sentence, from the loss of position and estimation outside of the prison, and not from any thought of forms of association within its walls. At the first, compulsory indiscriminate association with every class of criminals is doubtless felt as an augmentation of punishment by prisoners of the better sort; but with respect to any other object the alleged " degradation" loses its aspect of advantage when we consider that it debases the minds of its subjects, diminishes the remnants of self-respect, creates a feeling of community with the depraved, obscures for the class otherwise the most hopeful the prospect of any retrieval of lost character, and consequently that, as an element of any plan of reformation, it is suicidal.

Let me recapitulate, that in the adoption of the most efficacious known means of separation,—in the avoidance of compulsory intercommunication, and of the multitude of temptations resulting from proximity and the awakening of old associations,—in the maintenance of better relations, in some respects, between officers and prisoners,—in the exclusion of most of the inducements to mutiny and personal assaults,—in the application of special punishments or restraints for breaches of prison-rules,—in the option to abstain from such punishments, and to give free scope to spontaneous self-correction,—in the opportunities for the development of individual character,—in the protection given to the germs of good resolution,—in certain facilities for seasonable and appropriate instruction,—in the relation of labor to reformation,—in security against recognition, against enlargement of criminal acquaintanceships, and against the·general moral degradation caused by a forced public community in prison,—and, generally, in the more complete individualization of the discipline in its relations to the peculiarities of its subjects, cellular separation has advantages which are necessary consequences of that mode of imprisonment.

Let us here also avoid a fallacy. To prove the existence of an advantage is not to determine its quantity. It may appear that we have overrated or underrated the value to be assigned to the advantages just mentioned; but error in this respect will not destroy the fact that to some extent they exist. The friends of convict-separation maintain that they cannot easily be overrated, because they are the very objects for which the discipline of the old jails has been reformed, and because they are essential to the conception of either a just or a humane discipline. If we go beyond the mere punishment of an offender, or the procurement of the largest pecuniary benefit from his coerced labor, we must comprehend, as primary objects of our plans, the very particulars which cellular separation by its nature insures to us in larger measure than any other mode of confinement.

III. I believe it is not generally asserted anywhere, by intelligent observers, that, if the question between the leading systems were dependent only upon the considerations to which I have adverted, the plan of congregation would be preferable to its competitor. Conceding that *primâ facie* the favorable results claimed for separation are as already stated, it is ordinarily alleged that in practice they are neutralized or overbalanced by considerations of so peremptory a nature as to compel us to act upon other grounds and in subordination to other conditions than those presented by the mere separation of prisoners. Thus we are brought to the OBJECTIONS urged against the cellular method. They are—

1st. Its alleged influence upon bodily and mental health.

2d. Its alleged excess of pecuniary cost; and the inference thence that it is not "economical."

These are, in form, grave objections. Have we the means of ascertaining their real value? Let us take whatever precautions are suggested by the subject, and by our experience of its treatment heretofore; and with these safeguards we may find at least the conditions under which our inquiry is to be prosecuted.

In relation to HEALTH you will observe that there are two modes of investigation. First, we may assume certain universally conceded laws of hygiene, and then consider in connection with these the probable effects of any given prison regimen This is a theoretical procedure much used, though not always with satisfactory results. It ordinarily proceeds very few steps before it brings about a begging of the question on one or both sides; an unconscious introduction of tests and requisites, and kinds of evidence which are in reality contested.

It demands, too, the adoption of a criterion for the health which is proper for prisoners; and this is rarely adjusted beforehand in so clear a manner as to prevent misunderstanding at subsequent stages of the controversy. It has the further disadvantage that it depends in too great a degree upon the application of conclusions which were formed independently of prison experience, and which, therefore, do not embrace it under any law common to all the things to be compared. It brings general medical opinions, made outside of the circle of penitentiary life, to decide unqualifiedly questions arising within that circle.

The other mode of investigation professedly rests upon a collection of facts which are presumed to be decisive, and from which are inductively derived the ultimate conclusions of the inquirer. This mode claims to be "statistical," and therefore reliable.

What is really the thing sought? what is it, precisely, that we are to ascertain by either of these two modes? Is it whether or not the inmates of prisons exhibit or ought to exhibit the ruddy signs of robust health which we observe in the comfortable working classes, who lead regular lives and enjoy the benefits of home, with its ties and sympathies and regulating influences? Who ever saw such a state of things in any prison? who expects to see it? Is it whether or not the inmates of prisons exhibit equal manifestations of health with any one in particular of the classes of society from which any portion of the criminal population is derived? Who proposes so partial a test? What is the standard or type of health by which we are to learn the influence of any prison? Its inmates may have been taken from all grades of society, from every variety of occupation, from every kind of exposure to the privations of poverty, or from the abundance of vicious gratification. They may be from a healthful or a sickly district. Upon some of them food and comfortable lodging and temporary exemption from the anxieties of a precarious life will act as restoratives, and they will gain in health and spirits. Others will grow pale and attenuated from confinement and coarse fare. To some, labor will give wholesome exercise; upon others it will impose painful restraints. In every case there will appear good reason for a subdivision of our inquiry until we shall have made it agree with the subdivision of society out of doors.

Has any one accomplished this analysis, so that we may set out with an indisputable guide in our comparisons? It is not pretended that a shoemaker in the Eastern or Western Penitentiary will present to a casual observer the same aspect of health which is shown by a stonecutter working in the open air upon a public building at Columbus, or in a stonecutter's yard at Sing Sing or other congregate prison, or in the quarries, or in other out-of-door active employment. No one suggests that it is a part of the design of any system to keep the physical condition of convicts up to examples such as these; nor is it anywhere recommended that the State shall supply without stint the means of enjoyment to the extent of preventing in every individual a diminution of his strength or cheerfulness by his disciplinary seclusion. I insist upon this question: WHAT IS THE STANDARD? It is not very long since I read an essay written by an estimable gentleman, an intelligent friend of penal reform, and an expert medical practitioner of good repute among his professional brethren; and in that essay I found that an attempt was seriously made to ascertain the relations of cellular confinement to the health of a particular class of convicts, when the writer was confessedly ignorant of the separate medical statistics of that class in free life; and he even drew a conclusion from a comparison of the percentage of death and disease in that class in prison, with the percentage of many classes in the returns of the general census.

4

Will you take into consideration the number whose health has been improved in prison or has remained without change, as well as that of the inmates whose health is reported to have declined? For example, in one prison which was made the subject of special examination, there had been one hundred and twenty deaths; but it was found that in sixty-seven of these cases the convicts came into the prison in bad health. Further inquiry showed that of two hundred and twelve convicts who had been received in bad health one hundred and forty-five were discharged in improved health at the end of their terms of sentence, which ranged from one to ten years, and that all of them belonged to a class peculiarly liable to the diseases which are most commonly fatal in all prisons.

Will you register the condition of every individual and make your conclusion from the sum of such cases thus scrutinized?

Again, as each prison at the outset needed experience, which when acquired has lessened the number of deaths, at what stage of that experience shall we begin?

Who shall give to us the facts upon which we are to rely? If the medical officer of each prison, then during what series of years? It cannot have been forgotten that at Auburn, within a comparatively short period of each other, there were officially reported, by two physicians, two states of facts so inconsistent with each other as to make it certain that one of them was erroneous. In one year we were told that the bill of health compared so favorably with that of the separate penitentiaries as to confirm the judgment of the friends of associate labor; and soon afterwards it was asserted with great warmth of indignation by the second reporter that the health was bad, the insanity terrible, and that the truth had been concealed! In New Jersey we have had opposite conclusions officially reported by medical officers of the penitentiary of that State. Concede that it does not thence follow that our statistics are not to be derived from such officers, it nevertheless is evident that we cannot receive their statements without scrutiny.

Again, in what circumstances shall we require that the medical inspection shall be made, before agreeing to receive all reports upon the same footing of comparison? In the Eastern Penitentiary the change from a visiting to a resident physician immediately influenced the returns of disease. The more frequent and constant scrutiny, the concentration of attention, the better systematizing of observations, caused a minuteness and exactness of knowledge such as were impossible when the medical officer's tour of duty was confined to an interval taken from professional service in general practice.

Further, what allowance shall be made in relation to differences of professional opinion upon the necessary means of health?—for example, when one officer says, as in the Western Penitentiary, that the yards for stated exercise are not important, and another says, as at the Eastern Penitentiary, that out-of-door exercise is indispensable?

With respect to mental health, by what standard shall we judge of each prisoner upon his entrance into confinement, as well as during its continuance? It is familiar to all of us that the ideas of even medical officers are at variance upon this subject, and that quite opposite conclusions have been formed by them upon inspection of the same convicts.

The general notions which serve to guide conclusions in the community at large are inapplicable to a special class, and a special life, such as are embraced by our inquiry. The hereditary and other antecedents of criminals show an enormous predisposition to mental disturbance; but it has never yet been proved that, taking into view all classes of convicts, the system of Pennsylvania has been productive of worse results than are true of the congregate plan. The question is one of fact, not of supposition, nor of mere inference from premises which do not embrace the subject of the conclusion; and it is a consideration of great importance that the detection and report of symptoms of incipient insanity are more easy and certain under the peculiar inspection of individuals which is maintainable in the separate cells, than amid the distractions of the associate régime; which, as has long been known, cover from observation many indications of a decisive character. I do not hesitate to say that the standard of the medical officer

thus becomes more exact and appropriate; and that
it is more seasonably applied in separate than in
congregate prisons. Consequently our manifestation
of mental disturbance is more prompt and complete.
Yet, even with this feature of comparison against
us, no precise examination of facts has hitherto
sustained the popular preconception.

The cautions which have been suggested to you
under this head are not merely hypothetical: they
grow out of the history of prisons in the United
States. Perhaps with reference to no subject has
there been a greater abuse of so-called "statistics"
than has been practised in relation to prison health.
Figures have been employed as though they were
mere abstractions, instead of being representatives
of actual facts. Every rule of investigation proper
to the study of life and society has been violated
in turn. To European inquirers we are indebted for
the earliest satisfactory discussion of prison tables;
and the result should serve for a perpetual warning
against the distortions which characterized most of
the earlier controversial papers issued from the press
of our own country. There is not time now to cite
illustrations in detail; but a glance at the tables
arranged by Varrentrapp, or Julius, or Moreau
Christophe, will furnish good specimens of the statistical difficulties with which the system adopted
in Pennsylvania had formerly to contend. It is difficult to credit, yet it is true, that in one comparison
of the insanity at Charlestown with that in Philadelphia, there is no mention of the fact that legal pro-

vision had been made for the removal of insane persons from the prison at the former place, and that
no such provision existed at the latter; while in the
account given for the latter all the prisoners who
were insane at the time of their commitment, are
recorded without notice of their condition. Of
course, there is no reference to cases of persons with
hereditary predisposition to insanity whose condition was improved during confinement.

Not to dwell too long on this part of our subject,
let us ask, where precisely does the risk of ill health
begin? By necessary supposition for any fair comparison of the proper influences of each kind of
imprisonment, we must assume that the convicts
under each have sufficient food, clothing, lodging,
ventilation, work, and exercise for sanitary purposes.
It is then at the point of separation from one
another that we are to look for the mischief
charged upon the cellular method. Now, with
respect to this, we may obtain some aid to our
reflections by taking into view that measure of
the kind and extent of the alleged mischief which
is furnished by the remedies employed to prevent
it, and which, it is asserted, do prevent it in the
congregate prisons. Want of society is the cause.
What society is given? Not, indeed, such as implies
free and open conversation, or any such interchange
of thoughts and feelings as is understood elsewhere;
not that which is maintained between companions
who choose each other, and who have from that
fact a bond of sympathy, and the ordinary sti-

mulants and solaces of companionship; but professedly a chance aggregation under a prohibition of words, signs, and looks, which is enforced by punishments. What kind of mischief can be thus remedied? Is it not plain that its character has been misapprehended and its quantity exaggerated? Was such a method ever heard of in any other connection?

If we were to grant, however, that the advantage asserted is as great as it is claimed to be, there would remain the fatal objection that it is not obtained by any plan of discipline, by any thing formally recognized as a part of the congregate system. On the contrary, it is a consequence of a violation of the rules of the system. You prohibit intercommunication; the convicts have it in spite of you; and then you boast of the salutary hygienic effects of your regimen!

What, then, becomes of the grave question which involves the duty of the State not only to select the most efficacious discipline, but to provide with it suitable moral as well as physical safeguards? You meet this by a confessedly illicit society of convicts; for society in an effective sense it must be, to sustain the pretensions of the system. We demand that the society shall be that of honest people alone. If there is any extraordinary peril threatened by this plan, we demand that the State shall meet it, not by counteracting the motives to all penitentiary discipline, but by appropriate and legitimate means.

We deny the peril. Sustained by the most careful scrutiny of reports from prisons of every kind; corroborated by the judgment of a great majority of the experts who have most profoundly studied the question in Europe; with their recorded votes at Frankfort and at Brussels, and with the entire history of the subject before us, we unhesitatingly offer the records of the penitentiaries of Pennsylvania to general criticism. With the just qualifications to which all such records must be submitted, we fearlessly contest now, as with more restricted means did our predecessors, the advantage, in either a moral or a sanitary relation, which is claimed for the method of convict-association. In the course of the investigations which it is to be hoped the meetings of this Association will promote, there will be ample opportunity to verify and to weigh the appropriate evidence. My task is to define a position, not to sift a mass of proofs by a process of my own; but it is a grateful part of this task to pronounce anew the unwavering judgment of the Philadelphia Society, of which I stand here as a representative, while I also perform a duty assigned by this body. Since the foundation of the existing penal system of Pennsylvania, the members of the executive board of that society have been, by the law of the State, official visitors of its penitentiaries. That board is composed of men from every profession, who have been induced by humane feelings to visit our prisons and to "alleviate their miseries." Some of them enter upon their mission without previous knowledge of any part of our penal history. None of them come with a partisan spirit. All of them are keenly alive to the duty of promptly representing any real error in the administration of penal justice. They make many visits during each week to the cells in the Eastern Penitentiary, and to those in the Moyamensing prison. They find, as in all human institutions may be found, reasons to desire a more perfect representation of the divine wisdom and goodness; but from none of them, during these thirty years of visitation, has come a whisper of doubt that convicts should be separated one from another.

The officers appointed to administer the separate system in Pennsylvania, during these same thirty years, have been of a character to entitle them to every consideration due to general intelligence and to integrity of motive. Thus far, happily, the ideas of election by partisan votes, and of rotation in office, have not interfered with the steadiness of our administration. Whatever the imperfections of our method of discipline, it has never occurred to any of these gentlemen to question the soundness of our fundamental principle.

You are not asked to receive these testimonies as conclusive; but in the face of such facts, in view of the difficulties which have hitherto beset the general statistics of prisons, in view also of the very restricted nature of the alleviations proposed in the method of congregation, and of the fact that these are obtained in violation of its professed safeguards, it is not going very far to say that this Association has strong inducements to distrust preconceptions which have not been tested, and to pursue cautiously the road to its final decision.

This intimation has a further warrant in the fact that the investigations which are proposed by this Association are new to a considerable proportion of its members, at least in any sense which could render their judgment properly influential upon public opinion. Whether well or ill founded, that judgment cannot fail, according as it is in a wrong or a right direction, to multiply or to diminish the embarrassments to judicious legislation. A majority of the penitentiaries of this country are upon the congregate plan : the predispositions of the citizens here convened may be reasonably presumed to be against separation by day and by night, until the inquiries now to be opened shall justify a different bias. Let us quietly await the result of these inquiries, carefully conducted as they no doubt will be. Those of us who favor cellular separation will be always ready to give a fair scrutiny to the statistics which may be collected by your auxiliaries, and to submit to the conclusions which shall be sustained by them. Meantime, we feel quite safe in asserting that up to this moment there is not in the official reports from any prison, nor in a combination of reports from all the prisons, in the United States, any thing which upon an equal comparison will prove the alleged peculiarly dangerous effects of convict-separation upon the bodily or mental health of its subjects. After all that has been said upon the subject, it will probably surprise many of my fellow-members to learn that during the last ten years the mortality at the Eastern Penitentiary has been less than one per cent.

2. There remains to be noticed the argument upon the comparative "economy" of the two systems, with reference to which also there are some initial cautions which have been suggested partly by the natural elements of the question, and partly by experience during the controversy which has been maintained since the question first arose. What are we to understand by ECONOMY? Is it the pecuniary productiveness of a prison, without respect to the proper ends of imprisonment? Or are we still to regard that selection of ends, and the fitness of proposed means to them, which have already been stated as essential to any rational notion of public justice? In the latter case, we shall find that the tables of re-convictions, and the reports of moral and intellectual efficacy, must enter into our account. What views are entertained in this respect in Pennsylvania were fairly stated in the first report of the first warden of the Eastern Penitentiary: "I rejoice that it has never been the policy of the Legislature of this State to sacrifice the safety of the community and the welfare of the convict for apparent pecuniary gain: they have taken a higher, more dignified, and nobler ground; they have provided prisons where the reformation and improvement of the criminal and pro-

tection of society are grand objects; they have
provided that labor shall be furnished the convict in his cell, and not for the sordid purpose
of reimbursing to the Commonwealth the expense
of his maintenance." It is in consequence of such
views that up to this moment the visitors and
teachers of our convicts have daily withdrawn
them from their bodily labor for conversation and
instruction, thus necessarily diminishing the pecuniary fruits of their occupations, for the sake of
their intellectual and moral improvement. How
far in another direction official recommendations
have gone, may be seen in a late message of the
Governor of Wisconsin to the Legislature. "The
expenses of the State Prison," said his excellency,
"have been large for several years past, and are
necessarily increasing with the increasing number
of convicts. I would suggest to the Legislature the
propriety of leasing out, by a single contract, the
services of all the convicts, providing that they
shall be fed, clothed, and furnished with the
usual necessaries of life by the contractors, who
should also pay all expenses of guarding the prison,
and allow the State a reasonable compensation for
the services of the convicts. This system has been
adopted by some of the States, and proved, in its
results, mutually beneficial to both contracting
parties. The compensation for such services might
be applied toward the completion of the buildings
of the prison, and to other improvements connected
therewith"!

Again, where are we to begin our comparison of profits? When we visit the penitentiaries which stand as the best exponents in this country of the congregate system, we find in all of them great need of expenditure to put them on a just footing in relation to cost of construction. The size of the cells, the number of stories in the cell-blocks, the distribution of light and fresh air to the upper and the lower tiers of cells, the cleanliness, the freedom from all agencies within or without the cells deleterious to health or unfavorable to security, and other like topics, must, as has been before suggested, take precedence of any allegation of cheapness of maintenance. Above all, we have to dispose of the great question of the CONTRACT-SYSTEM OF LABOR, to which must be attributed a large percentage of the advantage claimed for associate workshops. What is the contract-system of labor? It is the hiring of the labor of convicts, at so much per day, to manufacturers, who send their own agents to instruct the workmen and to superintend the fabric. Of course, if the contract is worth having, it is worth competition. *Ceteris paribus*, he who bids highest will have the labor; and, by the ordinary consequence, he will expect to compensate himself, if possible, by the highest amount of productiveness of that labor. It is obvious that the first effect of this kind of management is to introduce between the prisoners and the officers who are intrusted with their discipline, a class of persons who, as has been remarked in another connection, are under no

other responsibility with respect to it than such as
concerns the custody of the convicts. Those persons are in continual intercourse with the convicts
throughout the day. Excepting in a few gross
instances, the officers have no voice, nor can they
have any, in the selection of them; because the
motive to their appointment is their skill in the
manufacture which they are to superintend, and
in obtaining the largest quantity of labor from the
prisoners. It is now rendered certain that these
conditions weaken the moral influence of the discipline; that they lead indirectly to favoritism towards
the best workers; that they create between the
professed objects of the law and the minds of the
prisoners an interval which prevents the contact
necessary for the best, or even for the ordinary,
fruits of a penitentiary life; that they expose the
prisoners to unnecessary irritations, manifested even
in homicidal assaults; that they lead to undue
exactions of labor, prejudicial alike to the respect
due to public law, and to the confidence of the
prisoners in the justice of its aims in their regard;
that they beget an indifference to the proper ends
of incarceration, and encourage the establishment of
a purely financial standard for the administration in
general. I say that these things are certain, because they are plainly attested in the official reports
of those prisons in which the contract-system prevails. I shall not repeat the strong phrases which
have been used by keepers, chaplains, and others
who have had practical experience of the system,

and who have denounced it with a force of language which, if employed here, might detract from your own confidence in the soberness of my remarks. I shall not for any present purpose cite the astounding fact disclosed by the investigation made by commissioners acting under the authority of the Legislature of New York,—commissioners before whom it was proved upon oath that agents were employed to attend the criminal courts, and to bribe convicts to say that they were accustomed to work at certain branches of manufacture, which in the Sing Sing prison had need of fresh recruits. It is sufficient to refer you to the annual reports from congregate prisons, and to those of the New York Prison Association, to furnish you with all the material requisite to sustain the assertions which have just been made to you.

Nor is this all of the difficulty which lies in your way. Will any one now venture the assertion, in the sense in which it will be understood in connection with my explanations, that any penitentiary in the United States is self-sustaining in a financial respect? The time has been when from many quarters we were officially told that prisons on the congregate system maintained themselves, or could be made to do so, and that on this account they deserved our preference. From no source did we receive more positive assertions on this head than from the penitentiaries of New York. Their annual reports never omitted the statement as a fact, that the public treasury was not burthened, or an expression of the confident hope that it would not thereafter be burthened, by any expense for the maintenance of convicts. Need I tell you how we of Pennsylvania were censured because we questioned the completeness of those reports; how the wounded honor of the wardens and their assistants resented our implied imputations upon their sincerity, and candor, and entire truthfulness; how our reasonable hesitation was employed to excite distrust of our fitness for the consideration of such matters in a manner sufficiently free from an obstinate partisan bias? You know what followed. The revelation which in later time has been made of the stupendous fraud practised upon the government and citizens of New York, and upon honest inquirers everywhere, is a memento not soon to become useless. Until, in every prison which is brought into comparison, we find officers willing to state not only every item of actual expenditure properly chargeable upon each year, but also every item of expenditure which ought to have been made in each year, it will be idle to pretend to any accuracy of conclusion. The mere difference in the mode of keeping the accounts occasions often very serious embarrassment. Some of us have seen a process by which, through the aid of a sinking fund, or a floating debt, a liability might be incurred in one year, amounting to many thousands of dollars, only the interest upon which, and the annual amount necessary to liquidate the principal in a long term of years, would appear in any one year's report.

Items of this kind have been cloaked under inappropriate heads of expenditure, in such a manner as to escape any scrutiny which should not be carried to the extent of an inspection of all the vouchers. Of course, to ascertain the real amount of expenditure for the first year of the series, one must wait to sum up the principal and interest account of all the years, extending, it may be, to ten, fifteen, or twenty. The Boston Society has informed us of another plan, which is stated to have been practised more than once, viz. that "a very favorable report has been made at the opening of the session of the legislature, concerning the ability of a prison to support itself, and before the close of the session a bill or resolve has been brought forward proposing to make an appropriation of several thousand dollars to the State prison for current expenses." These examples show the importance of the element of TIME; and they are, of course, introduced without the possibility of imputation upon those prison officers who have made their reckonings upon other principles. The report of the commissioners upon Sing Sing will give further information upon financial devices.

Again, it is not enough to take the books of one prison alone. The conveniences of the market; the character of the population from which the convicts come; the cost of food, and of the transportation of materials and fabrics; the restrictions imposed by legislation, and other qualifications, must be estimated. An officer of one penitentiary

5

informed me that, assuming the number 10 as the highest standard of capacity for workmen out of doors, he regarded 3½ as a full representation of the average capacity of the convicts admitted to the prison with which he was familiar; and that only about ten per cent. of the whole number of those convicts were skilled in trades practicable in that prison. These estimates were probably not exact; but they suggest two additional qualifications of our comparison. It is my own opinion that the official reports from which our figures must be taken are not yet sufficiently full and minute to warrant any positive general statement. The labors of this Association will, it is hoped, facilitate both the detection and the supply of deficiencies. In any case, it is proper to expect that due caution will be used in the selection of the prisons to be compared, and in the use of the particulars set forth in their tables.

Upon the whole, it may be safely denied that any manifestation has yet been made of the proper cost of a penitentiary on either plan of discipline; and our results are to be taken as in a measure accidental.

Before proceeding to my last principal subdivision, I beg to make a few suggestions having a general relation to what has been thus far said.

1. It is often objected that, notwithstanding the plausibility of the reasoning in favor of cellular separation, it makes no converts, no progress. This

objection is almost peculiar to this country; and it has a great effect upon persons unable or unwilling to investigate its value: yet the assumption of fact on which it rests is altogether erroneous. The separate system has made, to an extraordinary extent, both converts and progress; the causes which have anywhere retarded its advancement are well known, and they have no proper bearing upon the question of its real merits; and, were this not true, fresh inquirers would not be thereby dispensed from an examination of the evidence.[*]

Let us begin at the period when the establishment of the separate penitentiaries of Pennsylvania afforded the first opportunity to test upon a suitable scale the comparative merits of the rival systems. The opinion of almost the whole civilized world was against cellular separation. The cruel experiments in New York, Maine, Rhode Island, and Virginia had produced a general sentiment of condemnation, at home and abroad. The Auburn method was apparently working out the happiest results. The Boston Prison Society, whose annual reports were widely distributed, employed its resources in favor not only of congregate workshops, but even of the entire plan of construction adopted by New York.

---

* While these sheets are going through the press, the writer has felt himself at liberty to add to the text some paragraphs containing more precise details of the progress of the separate system abroad, than, from want of proper means of reference, could be given in his brief original summary. He regrets that a thorough revision of the essay cannot be made.

So influential were its representations, in conjunction with the pre-existing prejudice, that while the Eastern Penitentiary was in course of erection a vigorous controversy upon its plan arose in the legislature of Pennsylvania. The commissioners to whom its superintendence had been intrusted adhered to the separate method; the commissioners who had been appointed to revise the penal code and to adapt it to the new discipline, were so far moved as to pass beyond the limits of their function and to recommend an abandonment of the separate for the associate method. In the face of these discouragements, the persistent efforts of the Philadelphia Society obtained, at length, a confirmation by our legislature of its previously declared policy, and the Eastern Penitentiary went into opetation. Such was the aspect of affairs in 1829. Everywhere else, both public opinion and the action of governments were against cellular separation.

At the end of two years the French commissioners appeared among us, charged with a careful examination of the prisons of the United States. Let us take into view the sixteen years which intervened between their arrival in this country, and the congress at Brussels in 1847, at which the collective learning and practical experience of Europe were fully represented. We have no reason to question the value of such portions of our illustrations as are taken from abroad; for although some of the governments actively interested in penal reform are more arbitrary than our own, and hold

in less respect than ours the full liberty of the citi-
zen, yet the object of none of them was to augment
or insure the severity of punishment, but all of
them were seeking to render it more humane and
reformatory. The persons engaged in the discussions were men of established reputation for humanity as well as wisdom. Those discussions have
been printed, and are accessible to the public; and
difference of political institutions can be readily
ascertained to have exercised no influence unfavorable to the rights of individuals of any class.
Besides, more than ordinary respect is due to investigations, not begun and hastily completed within
a few weeks, as are too many of the inquiries of
legislative committees in America; not warped by
political partisanship, nor restricted by an undue
fear of unpopularity by reason of a liberal application of the public funds; but laboriously conducted
during a series of years, tested by all the resources
of systematic knowledge, and finally brought into
grave deliberation under the observation of the
world, and with·the certainty of criticism in every
community and by the statesmen of every government. The grounds upon which the congress rea-
soned were the same which have always been occupied by the friends of the separate system in the
United States; they were such as those which I
have imperfectly suggested.

It is a curious fact that this very multiplication of
experience and discussion has served in the United
States to discourage a general reference to it. The

keeper of a prison—especially if he has held his office during a large number of years—seems to himself to have acquired that superiority which in this country is always recognized as giving priority of claim to attention and confidence, viz. that of a practical man over a mere theorist. If he is a man of narrow acquirements, he distrusts every thing which he has not seen; and, full of the importance of his own observations, he refuses to listen to what he is told has been witnessed elsewhere. In proportion as the reasoning presented to him is generalized by the authority of facts in other quarters, he appears to become suspicious of speculation; that which to other men is a corroboration is to his mind an occasion for distrust; and he falls back within his little sphere of eyesight as though this were worth all the histories and reasoning of the world. It is in vain that you say to him that, if his experience is of such value, the experience of two or ten or a hundred keepers must have a proportionally greater weight; that, if the observation of a certain number of prisoners during a given number of years in one place is instructive, a like observation in many places must afford a better justification for an opinion. He replies that he is a "practical man;" and, like the warden of Massachusetts already mentioned, he values more highly his own experiment than the accumulated experience of ten thousand others. From him, as the manager of an important State institution, citizens and legislators derive the materials for their own opinions; and thus the wisdom of mankind is rejected for the judgments, and even the prejudices, of a single undisciplined mind. Against this extravagance it is not my purpose to offer any argument to this Association; but we shall do well to trace its effects, and to counteract them, elsewhere. Two different things were to be accomplished in any country which should seriously undertake the question of discipline with a view to practical measures. First, the government was to be convinced, and a permanent systematic policy was to be chosen, which must affect the entire scale of penalties, and, if conformed to the idea of convict-separation, must lead to a general reconstruction of prisons, large and small. If within a short space of time we had so far overcome the extraordinary difficulties in our way as to satisfy the legislature, or the executive, or even the commissioners, of only one of the enlightened governments of the world that our penitentiaries deserved their preference, this single fact might have been justly used as evidence of progress from the condition of affairs in the year 1829. In reality, within sixteen years, after scrupulous inquiry, not only had the commissioners of England, France, Prussia, Belgium, and other states, reported in favor of the superiority of the separate penitentiaries, but the governments of those countries, and of Sweden, Holland, and others, had all, upon unusually full public discussion, formally adopted the fundamental principle of separation as preferable to that of association of prisoners. When the European congress met at Frankfort in 1846, it was resolved by a large majority not only that persons detained for trial, but that convicts for long terms also, ought to be separated from one another by the cellular method.* So rapid a spread of a new opinion, against so many improbabilities, has certainly not many parallels.

The opinion was not one of merely abstract interest; it drew after it, as an immediate consequence, a large

---

* *Resolution 1st.* Separate or individual imprisonment should be applied to the accused (*aux prévenus et aux accusés*) in such manner that there cannot be any kind of communication, either amongst themselves or with other prisoners, except in cases in which, at the request of the prisoners themselves, the investigating officers (*les magistrats chargés de l'instruction*) may think proper to allow them certain relations within the limits determined by law.

Resolution 2d. Individual imprisonment shall be applied to the condemned in general, with the aggravations or mitigations required by the nature of the offences and of the sentences, the character and conduct of the prisoners, so that each prisoner be occupied with some useful labor, that he enjoy daily exercise in the open air, that he participate in the benefits of religious, moral, and literary instruction and in the exercises of religion, and that he receive regularly the visits of the minister of his religion, of the director, the physician, and the members of the committees of superintendence and patronage, independently of other visits which may be authorized by the regulations.

Resolution 3d. The preceding resolution shall be applied particularly to imprisonments of short duration.

Resolution 4th. Individual imprisonment shall be also applied to long terms of confinement, combining therewith all the progressive mitigations compatible with the maintenance of the principle of separation.—(*Proceedings of the Congress at Frankfort.*)

outlay of funds; and, as these could be obtained only upon legislative votes, it compelled each government to undertake the satisfying of the public mind. The thorough discussion of the whole subject led to one result which it is to be wished had been reached in Pennsylvania. It was perceived that any reform, to be thorough, must begin in those institutions which receive all classes of prisoners, and under whose influence, therefore, must come all the inmates of convict-prisons before their final incarceration. The course of procedure adopted was to inaugurate the new policy by a reconstitution of county prisons and houses of detention, and thus to prepare the way for an ultimate reconstruction of great penitentiaries. The whole work contemplated was enormous. For France alone 40,000 separate cells were required to be built; for Prussia, more than 13,000; and for other countries, a proportionate number. See what was accomplished within the sixteen years. In England a costly prison was erected, at Pentonville, in the outskirts of London, to serve as a model for the whole of the country; the government published plans for separate prisons of 12 to 500 cells; and more than 5000 such cells were built, in progress, and ordered. In France, as early as 1836, the Minister of the Interior issued a circular to the prefects of the Departments, informing them that in future the government would approve no plan for county prisons unless they should be such as to secure the absolute separation of prisoners; and ordering that the work on such of that class of prisons as were then in construction should be stopped until their plan could be changed. In 1841, another Minister of the Interior issued a circular with an atlas of plans for prisons of various sizes, from 12 to 160 cells. In 1847 there were already 23 of these prisons actually occupied, and many more were in course of construction. The government of Belgium, in 1844, presented to the legislative chambers a proposition for the introduction of the separate system into all the prisons of that country. Within three years thereafter, the foundations had been laid for more than 800 cells. Holland, by one sweeping provision, adopted the principle of the cellular system for all prisoners of each sex, and commenced the building of a separate prison of 212 cells. In Prussia, there were ordered five prisons on the cellular plan, some of which had been completed in 1847. One of these was to serve as a model; and all the plans of it were directed to be published. In Sweden, the legislature, by a formal vote, sanctioned the cellular system as the most rational and desirable, appropriated the large sum of more than one million of florins for the erection of new buildings in accordance with it, and began the actual construction of hundreds of cells. The enlightened king Oscar had, upon personal inspection, satisfied himself of the advantages of cellular separation, and in his own writings, as well as through the official papers of his ministers, urged its claims to preference.

Other proofs might be cited; but surely enough

has been said to show that, in consequence of a careful comparison of the separate and the congregate penitentiaries, the chief governments of Europe adopted a new policy, and began the reorganization of their prisons in accordance with the fundamental idea of cellular separation. Travellers from the United States do not, generally, notice prisons; and the few who occasionally turn from palaces and galleries of art to the contemplation of penal institutions, ordinarily fail to mark the fact that reform has been commenced in the minor prisons. They look at the great penitentiaries; and whenever these have not been remodelled they assume that the separate system has made no progress. Of course, it was not to be expected that a change so thorough, so extensive, so costly, as that which has been undertaken, should advance with an unabated rapidity such as that which characterized its outset. M. de Beaumont, so well known by his intelligent tour of observation in this country, appropriately remarked, in the legislature of France, that no government can commit such a folly as the pulling down of all its largest prisons at once, in order to build thirty at a time on a new plan, however advantageous. Besides, in other countries as well as our own there are fluctuations of interest, changes of administration, partisan opposition, foreign diversion, which retard the most clearly recognized improvements. It is not always possible to express immediately, through perfected machinery, our abstract conclusions. Legislation is not always at command.

Temporary biases of the public mind thwart or precipitate the execution of the wisest plans. The cogitations of the closet, even when these are legitimate fruits of a discreet consideration of acknowledged fundamental principles, make their way slowly to the general confidence of the mass of people. The population of no State is homogeneous. In Europe, as well as in the United States, public opinion is influential with the strongest governments. Thus, in Pennsylvania it has happened that the system of education, although it has received constant sanction and repeated aid from our legislature, is very unequal in its relation to different counties. In some of these the plan of the government for common schools has been satisfactorily executed; in others it has still very imperfect instruments. Thus, also, our penitentiary system remains unfinished, although from its foundation to this moment it has had the uninterrupted sanction of the legislature, through formal enactments needed for its interests, and annual appropriations in its favor. The county prisons, which should have been first reformed, continue for the most part at variance with the interests of the convict-discipline contemplated by our penal code. A special act passed for the control of their construction, which is ordered to be always so as to separate the prisoners, an act which requires all plans of gaols to be submitted to the Secretary of State and to have his approval before their final adoption, has not been so respected as even to secure the submission of the plans. Thirteen years ago our legislature provided for the official collection of materials for the discussion of every branch of our penal administration. That provision has remained almost a dead letter.

The European changes had to encounter the same difficulties which have presented themselves in the United States. With the design of informing the people, some of the governments abroad confined their first modifications to alterations of a wing of an old prison, or to the construction of prisons in the midst of the most intelligent portions of their population, in order to accustom the public mind to a consideration of the real features of the new plan. In 1848 broke out those domestic dissensions and foreign wars which have since absorbed the care of every European State. Considerations of finance, which would in any condition of things have been important enough to make progress slow, became more urgent under the pressure of a war-tax. They have been officially avowed as the motives for a suspension of progress in France and elsewhere. The concentration of attention, which had given to the separate system of discipline such aid in previous years, was broken up, and penal reform fell into the common routine of domestic affairs. It would be easy to trace the history of the penal institutions of each country to the present time; but my limits do not admit of so extensive an exposition. Suffice it to say that not one of the European governments which have adopted the policy of separation has since rescinded its resolutions in that regard; nor has there occurred any fact to shake the confidence of those who offered the cellular method to the world's criticism in the penitentiaries of Pennsylvania. In 1853, the government of Brazil sent an intelligent commissioner to examine the penitentiaries of the United States. He had the benefit of all the discussions and all the experience of Europe during the preceding twenty years to put him upon his guard. After a conscientious inspection of our prisons, he not only reported to his government, in a very decided manner, his preference of the separate prisons, but recommended that the work upon a House of Correction then in progress at Rio de Janeiro should be stopped, and the building be converted into a House of Detention; and that a new penitentiary should be constructed on the plan of cellular separation. So lately as within the last three years, or a little more, five commissioners appointed by the legislative authority of Frankfort-on-the-Maine to report upon the best plan for the structure and discipline of a new prison, after reviewing the history of prison systems in Europe and the United States, expressed, in the most unqualified phraseology, their conviction of the superiority of the plan of separation.

The model prison of England appears to require a few words to explain some peculiarities, and to bring into view some illustrations of the caution with which we should receive rumors of change, whether of opinion or practice. The original design of that prison was to receive convicts sentenced to transportation to the penal colonies. Eighteen months were to be passed in it for reformatory preparation; the remainder of the sentence was to be undergone in the convict-settlements. Good behavior in prison procured certain privileges in the colonies. Great care was employed to insure to each inmate a proper supply of the means of health. Experiments were tried upon food, the quality and quantity of which were changed from time to time; and each prisoner was weighed, to prove the effect of his diet. Partly in consequence of official reports, the term of confinement was reduced in 1848 to fifteen months, and in 1849 to twelve months; and in 1850 the Surveyor-General of Prisons, Lieutenant-Colonel Jebb, officially expressed the opinion that one year should be the average term. Here seems to be proof of the necessity of restricting cellular confinement to short terms. Yet two years later the chaplain of Pentonville, who had been chaplain of a hospital for the insane, and whose capability of discrimination had thus been favorably developed, published a volume which manifests unusual talent for his subject, in which he states that after the reduction of the term the bodily and mental disease largely increased. This seems to prove that the cases of ill health were not dependent upon the discipline. Looking further, however, we find that after the term had been shortened to one year a new class of prisoners were admitted, the average term of total sentence, including the period of transportation, was raised, and, instead of passing directly to the colonies, the convicts had to pass an intermediate period at the hulks or on the public works. We thus appear to lose the argument of the chaplain; but, continuing our investigation, it becomes known to us that the average both of disease and death among the convicts at Pentonville, whether before or after the change of term, was less than the percentage ascertained for persons of the same ages out of doors, less than that either at the hulks or on the public works, less than that of all the other prisons of England and Wales taken together, although nearly half of the committals to these are for less than two months: in fact, the mortality, allowing for pardons and removals for ill health, was nearly as small as at the Eastern Penitentiary, where it is less than one per cent. Were the comparison otherwise, regard being had to the mental impression produced by the prospect of transportation, and the excitements and irritations attendant upon the plan of rewards at Pentonville, and the fact that at the Eastern Penitentiary the sentences often exceed five years, and that many have been committed for terms ranging from ten to twelve, fifteen, and even twenty years, and that the average of all the sentences is much larger than that of Pentonville, we might safely regard the difficulties of the English prison as proceeding from some other cause than the influence of cellular separation. The resistance of the colonies to the shipment of criminals has raised a new problem for the home government, which is embarrassed by the inconveniences and risks of a continual liberation of large numbers of men of the dangerous classes within a small territory. It is to be hoped that from the exigence thus created will come some important suggestion with reference to the care of discharged convicts; but with respect to our topic it is enough to say that the bearing of such varieties of administration as are reported from Pentonville is upon the length of terms of sentence. Nobody in England officially questions the superior value of separate confinement for a reformatory discipline. Upon the relation of sentences to the plan of separation, something will be said hereafter.

In the United States it is true that there has hitherto been an unwillingness to adopt the cellular mode of imprisonment; but a consideration of the history of the subject ought to satisfy an impartial inquirer that no inference can be thence fairly drawn to overthrow, or even to qualify, the evidence presented by the penitentiaries of Pennsylvania. In this country it is not the usage to proceed by the systematic steps taken by foreign governments. Reforms are introduced into new prisons; but old establishments are not pulled down for the sake of inaugurating or extending a change of policy. Hence we have to await in each State the time when, without reference to a choice of penal systems, it becomes expedient to construct a new prison. When that time arrives, the subject is not thoroughly examined; nor is the evidence, after

being carefully discussed by the most competent
persons, presented in an elaborate communication
from one of the Secretaries of State. A bill is introduced by some one into the legislature. After
a few weeks, it is debated, and passed or rejected
mainly upon financial considerations. Perhaps now
and then a commissioner may be appointed to visit
the prisons of other States and to report the results
of his observation; but the choice of such an agent
is not, ordinarily, determined by his previous studies,
but rather by his general character as a man, and
by his claims upon the patronage of the appointing
power. To the people as well as to the legislature
the question of prison discipline is a distasteful and
unaccustomed subject; and, as each State is independent of the others, an inquiry into the penal
institutions of any of them is an investigation as
foreign as if it were carried into Canada or Brazil.
In these circumstances, a prejudice, or a feeling of
interest in pecuniary results, finds little to check it
from the side of patient discussion. It is an indisputable fact, moreover, that there has never ceased
to exist throughout this country, and there still
lingers in some portions of Europe, a very strong
prejudice against cellular separation. If this could
be regarded as the expression of a natural sentiment
in view of the whole case, it certainly ought to be
respected, and even to receive the treatment due to
an argument upon a question which involves the
rights of human beings under the administration
of public justice. It is not true, however, that the

resistance to the idea of convict-separation is directed against the system of Pennsylvania, but against the imaginary engine of cruelty which that system has been falsely assumed to be. The Boston Society's reports, and the occasional apparent confirmation which these have received from unqualified observers, have continued to furnish topics of objection where no answer could be seasonably made. The surprise expressed by most persons, who visit for the first time the Eastern Penitentiary, at the provision there made for the health and comfort and improvement of prisoners, attests the strength of the prepossession against the discipline of the place. Perhaps the best illustration of this kind of influence is to be found in the report made by a distinguished English author of his visit to that penitentiary. Availing himself of his reputation and of his skill with the pen, he published a narrative in which the horrors of isolation were depicted in such a manner as must shock every sensitive mind. His myriads of readers in this country and Great Britain received as verity what came upon testimony apparently so trustworthy; and thus Mr. Dickens became the expositor of the value of cellular separation in its relations to health, reformation, and the sanctity of penal justice. Those persons who were most familiar with the real state of things were confounded by the almost inconceivable misstatements of the published report; and, as the convicts with whom Mr. Dickens held his conversations had been noted by the warden of the penitentiary, the British consul, a gentleman of liberal education, was invited to follow the romancer's route from cell to cell. The investigation might be termed ludicrous, were it not for the important relations which it had to public opinion upon a momentous public question. Not one of all the list of examples stated by Mr. Dickens was found to justify, in any degree, his report; and the climax of confutation was given when the consul stated that the three young women who had chiefly enlisted the general sympathy were mulattoes, decoys of a low brothel; and that all of them had been consciously benefited by their incarceration.*

Of the mode in which investigations authorized by the legislature of one of our States may be conducted so as to reach a conclusion with entire ignorance of the elements of the question proposed, an example may be offered in the case of Missouri, in whose behalf a commissioner visited the principal penitentiaries of the Atlantic States in 1852. Placing himself in communication with the managing agent of the Boston Society, he collected the stale, often confuted "statistics" and descriptions of that society's reports, and repeated them as though verified by his own observation. He totally misrepresented the regulations and actual practice of the peniten-

---

* The writer thinks it expedient to give, in an appendix, the narrative of Mr. Dickens and the comments of the consul, Mr. Peter. The narrative was reprinted as EVIDENCE in the eighteenth annual report of the Boston Society!

tiaries of Pennsylvania, though he had visited these. He altogether perverted the evidence of cost, even to the extent of asserting that the erection of a congregate prison was "not half so expensive" as that of one on the separate plan; and he held out the mythical bait that a congregate prison, "after paying for itself during the first few years of its existence, will thereafter yield annually a handsome revenue to the State." A more worthless, a more pernicious, document of its kind could scarcely be officially presented to any government. Yet it was a formal report by a selected commissioner, made after months of travel and inquiry. How were the members of the legislature to know that it was false even as to most obvious facts?

In New Jersey, where a penitentiary designed for convict-separation was erected upon plans furnished by the architect who constructed the large prisons of Pennsylvania, it was hoped that an opportunity would be gained for a comparison of results, and that the statistics contributed would give an advantage proportionate to the additional number of persons who were to come under the influence of cellular confinement. The two States being adjacent, and the prison of one being modelled after those of the other, we ought to have had at least an approximation to sameness of management in all of them. In the tables of foreign writers, the penitentiary of New Jersey has been usually placed upon the same footing as the Eastern and Western Penitentiaries, for the purpose of comparison with the congregate-prisons. If now it be announced, without explanation, that, an enlargement of the penitentiary of New Jersey having become necessary, the government of that State has abandoned the plan of separation and has ordered a new wing to be adapted to associate labor, the citizens of other States will reasonably conclude that here at least is direct evidence against the cellular method, and that the proposed change is a legitimate result of experience. The officers who have had the oversight of the prison will be appealed to as witnesses of the practical superiority of the congregate plan. From our own friends will seem to arise conclusive proof against us. It is nevertheless true, as there has been previous occasion to remark, that during many years the penitentiary of New Jersey has not been regarded by the Philadelphia Society as an example of its own plans. No officers of that penitentiary who have been in function during those years can be properly said to have had experience in it either of the separate or of the congregate system; and, whatever respect may be due to their general intelligence and character, the testimony of none of them as to either of these systems would be accepted by any practised inquirer. When the augmented population of the State began to crowd the cells, and larger accommodation was needed. it was alleged that financial economy would be promoted by the introduction of associate labor. The reports of congregate prisons, taken in the unqualified form in which it is customary to throw receipts and expenses into their general annual statements of accounts, were offered to the members of the legislature as examples of what might be done to lessen taxation. The calculations of the officers of the prison, to show what might be gained by the common workshops, strengthened the expectations based upon the reports. There was no general review of the discussions of experts elsewhere—no thorough scrutiny of the returns of health—not even a proper consideration of that main stay of the allegations of profit, the contract-system of labor. Upon the irrelevant experience of a misused penitentiary, and upon a promise of pecuniary advantage, the legislature authorized the construction of a cell block and shops for associate labor.

Within the last five years a former warden of the prison officially reported, "We have enjoyed extraordinary good health." "The discipline of the prison is in a wholesome condition. No serious offences, and but one case of insubordination has occurred. The rules to regulate the deportment of prisoners are characterized by a mildness only practicable under the separate system." "So long as the reformation of the convict is to be considered a controlling purpose of the discipline to which he is subjected, his labor and the profits resulting therefrom must be made a secondary consideration." At the same time, a special committee of the legislature reported that the greatest evil which they could discover was "the political character of the prison, and the mutations of government to which it is liable from the frequent changes of party. The great qualification which seems now to be taken into consideration is the peculiar tenets of the keeper, and not the fitness or ability which is requisite for an office in which so much depends upon its executive." The committee asserted positively that in New Jersey "the separate system had not been fairly tested;" and they said, "After a careful examination of the two systems which have been adopted in this country, the separate and the congregate, or the silent, they have arrived at the conclusion that the preference in all respects is to be awarded to the former." What has happened during these five years to justify a change of policy? Is it not clear that the case offers only another example of that which the Brazilian commissioner noticed as being at once foreign to the purpose of punishment, and as characteristic of Americans, viz., a preference of what is thought to cost the fewest dollars?

There is another aspect of the question of progress which ought not to be omitted. During the sixteen years taken for our term of observation, there may be seen a gradual diminution of the claims of the congregate system to those grounds of absolute confidence which were once thought to be indisputably occupied. Its statistics of health have been confuted; its boast of "profit" has been disproved by the accumulation of years of account, and even the semblance of profit has been purchased by an unjust system of labor; its alleged success in

the maintenance of non-intercourse has been shown
to be an impossibility: so that at the very time at
which the system of Pennsylvania, having grown
in the confidence of its early friends, was achieving
a climax of triumph in the European congress, we
find that in New York an assemblage of wise and
experienced citizens, led by one of the Inspectors of
Prisons for that State, were deliberately expressing
the opinion that something better than the plan of
Auburn was needed.

2. This reference to the New York Prison Association suggests some further cautions. It is currently believed that so wedded is the Philadelphia
Society to the idea of cellular separation that no
quantity or strength of evidence can affect in the
slightest degree the partisan bias of its members.
Yet it ought to be as currently remarked that that
society is at least entitled to demand a correct appreciation of its real position, and an acknowledg-
ment of the fact that, excepting in relation to the
single feature of separation of convicts one from
another, it is always occupied with the investigation of possibilities of improvement. In this connection we perceive the value of a clear definition
of the word "system." When the French architects made in their plans a special arrangement of
the yards for exercise, and contrived such a dispo-
sition of their chapels as would allow the prisoners
to witness the celebration of mass, the phrase
"French system" was immediately employed, although nothing was new except mechanical changes

intended to meet local religious opinions, or to supply for out-of-door exercise greater conveniences than had been attained at the pioneer prison. In such a sense, every alteration must justify the allegation of a change of system; and in the same sense the friends of cellular separation desire every improvement that can be reasonably commended to the attention of governments. They have never treated the Eastern Penitentiary as a model prison to any greater extent than was warranted by the fact that it offered the largest and most conveniently situated illustration of the plan of convict-separation. Every one of its seven blocks of cells exhibits improvement; and between the first and the seventh are differences as great as are usually observable between prisons of different States. The large county prisons erected in Pennsylvania at later dates, and designed in part for convicts, have derived advantage from foreign plans; and in some respects their details are superior to those of the penitentiaries. Thus, too, with regard to diet, exercise, special punishments, the apportionment of sentences, and the supply of labor, the Philadelphia Society has unremittingly sought counsel. A few years ago it sent a medical commission to inspect the most important congregate prisons of this country, in order to procure ampler means of judgment upon the sanitary regulations proper to cellular confinement. It has repeatedly applied to the government for a more complete manifestation of the apportionment of penalties for the various classes of

crime; and it has long since represented its conviction that the sentences passed under our code
are frequently too long for the character of the
confinement to which they subject offenders. Not,
indeed, that the extreme conclusion of the British
Surveyor-General, counteracted as this is by nu-
merous reports, official and others, and by our own
experience, has ever been adopted in Pennsylvania;
not that the average term of imprisonment has been
ascertained to be greatly in excess; but there have
been imposed upon certain of the prisoners dispro-
portionate terms. These affect both the discipline
and the health of our penitentiaries: yet, in spite
of this disadvantage, our statistics are favorable.

In the same spirit we have always desired that
where the congregate plan is preferred it shall have
all the instrumentalities which are needed for its
humane administration; but when its pecuniary
reports are claimed as an advantage, we cannot
but censure the labor-system which sustains them.
When it assumes to prevent evil intercourse by
the discipline of its shops, we cannot but accept
the proofs which come to us from every prison in
this country and in Europe in which such prevention has been attempted, that it is imprac-
ticable.

In the hope of avoiding the difficulties which
have involved in prolonged controversy the methods
of congregation and separation, it has been repeatedly suggested that there may be devised a
system which shall have the advantages, while it

avoids the peculiar evils, attributed to each. On the continent of Europe the full recognition of cellular separation was preceded in some countries by experiments having in view this mixture of features. In one locality, a separate and an associate prison were placed side by side; in another, a portion of the same building was given to the trial of each form of discipline, and convicts were classified and distributed according to their supposed character; sometimes the cellular confinement was used as an introduction to the privileges of the associate shops, which were treated as places of promotion, and unruly subjects were remanded from them to the separate cells. In England, as we have seen, the cells were intended to be places of preparation for associate labor on the public works and in the penal colonies. In the endeavor to make classification serve the requirements of safe custody and discipline, the subdivisions have been carried to as many as fifteen, twenty, even nearly forty, classes; and everywhere they have failed, unless in relation to indications so gross as to be valueless for our question. Men may easily be grouped according to age, crime charged, time of sentence, health, previous profession, behavior in prison, skill in labor; but for any nice discrimination for discipline the means have never yet been discovered. I do not pretend to set bounds to human invention, nor to assume that the thoughts which are familiar to us will always constitute our stock of knowledge; but it is not easy to see the wisdom of repeating trials which have already failed in the hands of expert officers of each sex. At all events, whenever classification is offered as a safeguard, let us hope that our legislatures will at least inquire what number and kind of persons have already subjected it to patient observation during many years, and what account they give of the actual results.

The New York Prison Association, unwilling to sanction the cellular method for all convicts, or for the entire term of their sentences, yet perceiving some of its advantages, sought to encourage that union and development of particular features which have been so often suggested abroad. The contract-system was to be proscribed; the prisoners were to be classed, first to assign some of them to separate cells, and others to the associate wing; next to group the prisoners in the latter, according to certain types of character to be detected by the scrutiny of the officers. Intercommunication was not to be prohibited to the extent theretofore considered proper; but some reliance was to be placed upon the success of official scrutiny and supervision; and a regulated intercourse among the convicts was to aid the reforming efforts of the administration, and to train repentant minds for the social opportunities which awaited their discharge. This is an attractive picture of a penitentiary administration. Is it warranted by experience? I freely submit it to you for comparison with the records of prisons at home and abroad during the last thirty years. Its rejection of the contract-system, and its promise of suitable officers, will inevitably deprive it of the financial superiority claimed by the congregate method.

For my own part, were it not that the cost of reconstruction embarrasses so greatly every proposal for change of discipline, I would be glad to see the suggestions of the New York Prison Association brought to trial under the eyes of its intelligent and humane members. It would be an important end gained to break the force of habit and to stir the public mind to fresh thoughts, and it would be a step towards the ultimate separation of every convict from every other convict; but considering the length of the period that must ensue before any State would make a second change of a radical character, I prefer to avoid the intermediate loss, and to ask that the next experiment be made with a principle which, with all the defects of its administration in Pennsylvania, has never suggested there the expediency of seeking something better.

It will be understood that no State is asked to tear down its great prisons for the sake of reconstructing them upon the plan of separation; but every State may adopt a policy, and work towards its execution as time brings motives for action. Old penitentiaries need alterations; increase of population demands new blocks of cells; occasionally in a new State the opportunity is offered to lay the foundation of a system. It is thus that all countries slowly advance towards the consummation of the plans of their governments.

It is with great regret, Mr. President, that I have found myself obliged to present in so imperfect a manner some of the considerations which still sustain the friends of convict-separation in their opposition to the companionship of criminals. I proceed with more alacrity to the notice of the last subdivision of my reflections—that which embraces the topics upon which all parties may unite without conflict with their respective conclusions upon the question of cellular confinement for convicts.

IV. After what has been said, you will not be surprised if I put in the foremost place the COLLECTION OF EVIDENCE. The accomplishment of this alone will give to the Association a very extensive, and at times a very laborious, occupation. We have first to determine what particulars of evidence are needed. The number of these is great; and their ascertainment depends upon a careful examination both of the questions to be decided, and of the probable sources of error. We have then to agree upon a uniform method in which the particulars shall be reported, and to provide securities against errors of observation and errors consequent upon misconception of our objects. Then we have to obtain the practical adoption of our method by the officers who are to record and publish the particulars. This will require, in many places, fresh legislation; and as some of our requisites will occasion additional labor, or expense, or both, it will be necessary to use the means appropriate for convincing members of the legislatures, executive officers, and a portion of the community at large, that our purpose is such as to entitle it to aid from the authorities of each State. When all this has been done, it will be our duty in each State to see that the official records be kept in conformity with the general plan, and that the official reports comprehend all the necessary items duly ascertained. It will be our business to awaken the public interest in such a manner as to insure the steady maintenance of official action; and we ought to establish a permanent collection of statistical documents to which reference may be conveniently made during our joint deliberations. In no State has any portion of this large undertaking been satisfactorily executed. In Pennsylvania, the law obtained in 1847, upon the memorial of the Prison Society of Philadelphia, was all that could be accomplished at that time, even in the way of formal enactment; and it does not embrace all of the desirable details of our penal administration. Its provisions will, however, serve to illustrate the character of the inquiries which we need. As before noticed, it has not been enforced; nor can it be, until further steps shall have been taken to satisfy the government that official duty requires its full execution.

As our penal institutions are designed for every agency of public justice, from the detection of an offender to the completion of the disciplinary infliction awarded to him, it is evident that police, the

functions of committing magistrates, primary deten-
tions, courts and juries, as well as prisons, are
proper objects for the deliberation and efforts of
this Association. With respect to all of these the
public mind needs information; all of them need
reform.

The sources of crime, and the philosophy of the
dangerous classes; the definition of new crimes; the
relation of judicial sentences to the various classes
of crime, and to the fundamental ideas of penal
discipline; the extent of the influence of that discipline, and the ends to which it ought to be directed;
the peculiarities of treatment due to persons confined for life; the care of discharged convicts, and
kindred topics, await your discussion. With respect to all of these the friends of both systems
of imprisonment may co-operate without difficulty.

One topic I notice specially, because a peculiar
misconception exists in many minds in relation to it.
In the treatment of the question of separation by
day and by night, prisons for detention before trial
have too often been classed with prisons for convicts, as though the former were liable to the same
objections, and depended upon the same reasons,
which influence our conclusions respecting the
latter. In Europe the distinction has been clearly
marked. At the Frankfort convention in 1846,
where there were some voices raised against the
cellular separation of convicts, there was unanimity
in favor of that plan of confinement in the prisons

7

used before trial. I believe that only one negative vote was given. The New York Prison Association early recognized the distinction; as may be seen in its reports. In Great Britain, so confused at one time was public opinion on this subject that objection was made in Parliament to the separate imprisonment of unconvicted persons, on the ground that it was cruelty to the innocent! About twelve years ago, Lord Nugent moved to repeal so much of the act of the 2 & 3 Victoria as gave power to magistrates to inflict separate imprisonment upon persons committed for trial; a power which appeared to him to be "inconsistent with every principle of general justice and with the whole spirit of the criminal code." The masterly reply of Sir George Gray induced the mover to withdraw his proposition, which he did in accordance with the general feeling of the House. The pith of the argument is summed up by the chaplain to the Preston House of Correction :—"Should it be objected that to separate the untried is to punish them, and that punishment must not be inflicted until guilt is proved, I would reply that a prisoner committed for trial must be either guilty or innocent—an adept in crime or a novice. If the former, separation is no injustice to him; for he has no right to be placed among those whom he would contaminate. If, on the other hand, the newly committed prisoner be innocent or unused to crime, he has a right to be protected from influences which would inflict upon him a horrible and irreparable injury."

Since, then, the jails for detention before trial
are quite free from the objections made against tne
separate imprisonment of convicts; objections which
relate mainly to health and economy during long
periods of incarceration, what is to hinder a joint
effort for such a reform of the county prisons of the
United States as will answer the ends of public
justice and individual safety? Allowing the question as to convicts to be discussed on its peculiar
reasons, we may unite in a plan for the improvement of those institutions which precede in order
of use, and which materially influence, our convict
discipline.

I should be glad to believe, Mr. President, that
the labors of this Association will contribute to the
public good in the respects in which that is dependent upon the topics which have been mentioned to
you. With zeal tempered by rational caution, with
energy, with perseverance, with patience, with readi-
ness to receive the instruction of facts, and with
harmonious co-operation, we may procure such a
concentration of social forces upon the great work
before us as will insure its accomplishment. Let
no apprehension of temporary popular opposition
sway us from the steadfast maintenance of whatever policy we know to be the best for the State.
If considerations of cost or of time interfere with
the immediate realization of our plans, let us not,
while yielding to the inevitable constraint of present
circumstances, suspend our assertion of the princi-

ples upon which the penal jurisprudence of the country ought ultimately to rest. It is by adherence to these through every vicissitude of means, of opportunity, and of resistance, that we are to gain at last the public confidence, and with it the crown of success.

APPENDIX.

Mr. Dickens' Report of his Visit to the Eastern Penitentiary.

"In the outskirts, stands a great prison, called the Eastern Penitentiary: conducted on a plan peculiar to the state of Pennsylvania. The system here, is rigid, strict, and hopeless solitary confinement. I believe it, in its effects, to be cruel and wrong.

"In its intention, I am well convinced that it is kind, humane, and meant for reformation; but I am persuaded that those who devised this system of Prison Discipline, and those benevolent gentlemen who carry it into execution, do not know what it is that they are doing. I believe that very few men are capable of estimating the immense amount of torture and agony which this dreadful punishment, prolonged for years, inflicts upon the sufferers; and in guessing at it myself, and in reasoning from what I have seen written upon their faces, and what to my certain knowledge they feel within, I am only the more convinced that there is a depth of terrible endurance in it which none but the sufferers themselves can fathom, and which no man has a right to inflict upon his fellow creature. I hold this slow and daily tampering with the mysteries of the brain, to be immeasurably worse than any torture of the body: and because its ghastly signs and tokens are not so palpable to the eye and sense of touch as scars upon the flesh; because its wounds are not upon the surface, and it extorts few cries that human ears can hear; therefore I the more denounce it, as a secret punishment which slumbering humanity is not roused up to stay. I hesitated once, debating with myself, whether, if I had the power of saying 'Yes' or 'No,' I would allow it to be tried

97

in certain cases, where the terms of imprisonment were short; but now, I solemnly declare, that with no rewards or honours could I walk a happy man beneath the open sky by day, or lie me down upon my bed at night, with the consciousness that one human creature, for any length of time, no matter what, lay suffering this unknown punishment in his silent cell, and I the cause, or I consenting to it in the least degree.

"I was accompanied to this prison by two gentlemen officially connected with its management, and passed the day in going from cell to cell, and talking with the inmates. Every facility was afforded me, that the utmost courtesy could suggest. Nothing was concealed or hidden from my view, and every piece of information that I sought, was openly and frankly given. The perfect order of the building cannot be praised too highly, and of the excellent motives of all who are immediately concerned in the administration of the system, there can be no kind of question.

"Between the body of the prison and the outer wall, there is a spacious garden. Entering it, by a wicket in the massive gate, we pursued the path before us to its other termination, and passed into a large chamber, from which seven long passages radiate.

"On either side of each, is a long, long row of low cell doors, with a certain number over every one. Above, a gallery of cells like those below, except that they have no narrow yard attached (as those in the ground tier have), and are somewhat smaller. The possession of two of these, is supposed to compensate for the absence of so much air and exercise as can be had in the dull strip attached to each of the others, in an hour's time every day; and therefore every prisoner in this upper story has two cells, adjoining and communicating with, each other.

"Standing at the central point, and looking down these dreary passages, the dull repose and quiet that prevails, is awful. Occasionally there is a drowsy sound from some lone weaver's shuttle, or shoemaker's last, but it is stifled by the thick walls and heavy dungeon-door, and only serves to make the general stillness more profound. Over the head and face of every prisoner who comes into this melancholy house, a black hood is drawn; and in this dark shroud, an emblem of the curtain dropped between him and the living world, he is led to the cell from which he never again comes forth, until his whole term of imprisonment has expired. He never hears of wife or children; home or friends; the life or death of any single creature. He sees the prison officers, but with that exception he never looks upon a human countenance or hears a human voice. He is a man buried alive; to be dug out in the slow round of years; and in the mean time dead to every thing but torturing anxieties and horrible despair.

" His name, and crime, and term of suffering, are unknown, even to the officer who delivers him his daily food. There is a number over his cell-door, and in a book of which the governor of the prison has one copy, and the moral instructor another : this is the index to his history. Beyond these pages the prison has no record of his existence : and though he live to be in the same cell ten weary years, he has no means of knowing, down to the very last hour, in what part of the building it is situated; what kind of men there are about him; whether in the long winter nights there are living people near, or he is in some lonely corner of the great jail, with walls, and passages, and iron doors, between him and the nearest sharer in its solitary horrors.

"Every cell has double doors: the outer one of sturdy oak, the other of grated iron, wherein there is a trap through which his food is handed. He has a Bible, and a slate and pencil, and, under certain restrictions, has sometimes other books, provided for the purpose, and pen, and ink, and paper. His razor, plate, and can, and basin, hang upon the wall, or shine upon the little shelf. Fresh water is laid on in every cell, and he can draw it at his pleasure. During the day, his bedstead turns up against the wall, and leaves more space for him to work in. His loom, or bench, or wheel, is there; and there he labours, sleeps and wakes, and counts the seasons as they change, and grows old.

* " The first man I saw, was seated at his loom, at work. He had been there, six years, and was to remain, I think, three more. He had been convicted as a receiver of stolen goods, but even after

---

[* These cases are referred to by numbers in the subjoined letter of the British consul. Mr. Peter.]

this long imprisonment, denied his guilt, and said he had been hardly dealt by. It was his second offence.

"He stopped his work when we went in, took off his spectacles, and answered freely to every thing that was said to him, but always with a strange kind of pause first, and in a low, thoughtful voice. He wore a paper hat of his own making, and was pleased to have it noticed and commended. He had very ingeniously manufactured a sort of Dutch clock from some disregarded odds and ends ; and his vinegar-bottle served for the pendulum. Seeing me interested in this contrivance, he looked up at it with a great deal of pride, and said that he had been thinking of improving it, and that he hoped the hammer and a little piece of broken glass beside it 'would play music before long.' He had extracted some colours from the yarn with which he worked, and painted a few poor figures on the wall. One, of a female, over the door, he called 'The Lady of the Lake.'

"He smiled as I looked at these contrivances to wile away the time ; but when I looked from them to him, I saw that his lip trembled, and could have counted the beating of his heart. I forget how it came about, but some allusion was made to his having a wife. He shook his head at the word, turned aside, and covered his face with his hands.

"'But you are resigned now!' said one of the gentlemen after a short pause, during which he had resumed his former manner. He answered with a sigh, that seemed quite reckless in its hopelessness, 'Oh yes, oh yes! I am resigned to it.' 'And are a better man, you think?' 'Well, I hope so : I'm sure I hope I may be.' 'And time goes pretty quickly?' 'Time is very long, gentlemen, within these four walls!'

"He gazed about him—Heaven only knows how wearily!—as he said these words ; and in the act of doing so, fell into a strange stare as if he had forgotten something. A moment afterwards he sighed heavily, put on his spectacles, and went about his work again.

"In another cell, there was a German, sentenced to five years' imprisonment for larceny, two of which had just expired. With colours procured in the same manner, he had painted every inch of the walls and ceiling quite beautifully. He had laid out the few feet of ground, behind, with exquisite neatness, and had made a little bed in the centre, that looked by the bye like a grave. The taste and ingenuity he had displayed in every thing were most extraordinary; and yet a more dejected, heart-broken, wretched creature, it would be difficult to imagine. I never saw such a picture of forlorn affliction and distress of mind. My heart bled for him; and when the tears ran down his cheeks, and he took one of the visitors aside, to ask, with his trembling hands nervously clutching at his coat to detain him, whether there was no hope of his dismal sentence being commuted, the spectacle was really too painful to witness. I never saw or heard of any kind of misery that impressed me more than the wretchedness of this man.

"In the third cell, was a tall strong black, a burglar, working at his proper trade of making screws and the like. His time was nearly out. He was not only a very dexterous thief, but was notorious for his boldness and hardihood, and for the number of his previous convictions. He entertained us with a long account of his achievements, which he narrated with such infinite relish, that he actually seemed to lick his lips as he told us racy anecdotes of stolen plate, and of old ladies whom he had watched as they sat at windows in silver spectacles (he had plainly had an eye to their metal even from the other side of the street), and had afterwards robbed. This fellow, upon the slightest encouragement, would have mingled with his professional recollections the most detestable cant; but I am very much 'mistaken if he could have surpassed the unmitigated hypocrisy with which he declared that he blessed the day on which he came into that prison, and that he never would commit another robbery as long as he lived.

"There was one man who was allowed, as an indulgence, to keep rabbits. His room having rather a close smell in consequence, they called to him at the door to come out into the passage. He complied of course, and stood shading his haggard face in the unwonted sunlight of the great window, looking as wan and unearthly as if he had been summoned from the grave. He had a white rabbit in his breast; and when the little creature, getting down upon the ground, stole back into the cell, and he, being dismissed, crept timidly after it, I thought it would have been very hard to say in what respect the man was the nobler animal of the two.

"There was an English thief, who had been there but a few days out of seven years: a villanous, low-browed, thin-lipped fellow, with a white face; who had as yet no relish for visitors, and who, but for the additional penalty, would have gladly stabbed me with his shoemaker's knife. There was another German who had entered the jail but yesterday, and who started from his bed when we looked in, and pleaded, in his broken English, very hard for work. There was a poet who, after doing two days' work in every four-and-twenty hours, one for himself and one for the prison, wrote verses about ships, (he was by trade a mariner,) and 'the maddening wine-cup,' and his friends at home. There were very many of them. Some reddened at the sight of visitors, and some turned very pale. Some two or three had prisoner nurses with them, for they were very sick; and one, a fat old negro whose leg had been taken off within the jail, had for his attendant a classical scholar and an accomplished surgeon, himself a prisoner likewise. Sitting upon the stairs, engaged in some slight work, was a pretty coloured boy. 'Is there no refuge for young criminals in Philadelphia, then?' said I. 'Yes, but only for white children.' Noble aristocracy in crime!*

"There was a sailor who had been there upwards of eleven years, and who in a few months' time would be free. Eleven years of solitary confinement!

"'I am very glad to hear your time is nearly out.' What does he say? Nothing. Why does he stare at his hands, and pick the flesh upon his fingers, and raise his eyes for an instant, every now and then, to those bare walls which have seen his head turn grey? It is a way he has sometimes.

"Does he never look men in the face, and does he always pluck at those hands of his, as though he were bent on parting skin and bone? It is his humour: nothing more.

[* There has been since completed and occupied a House of Refuge for coloured children at Philadelphia.]

" It is his humour too, to say that he does not look forward to going out; that he is not glad the time is drawing near; that he did look forward to it once, but that was very long ago ; that he has lost all care for every thing. Is it his humour to be a helpless, crushed, and broken man. And, Heaven be his witness that he has his humour thoroughly gratified !

" There were three young women in adjoining cells, all convicted at the same time of a conspiracy to rob their prosecutor. In the silence and solitude of their lives, they had grown to be quite beautiful. Their looks were very sad, and might have moved the sternest visitor to tears, but not to that kind of sorrow which the contemplation of the men, awakens. One was a young girl ; not twenty, as I recollect; whose snow-white room was hung with the work of some former prisoner, and upon whose downcast face the sun in all its splendour shone down through the high chink in the wall, where one narrow strip of bright blue sky was visible. She was very penitent and quiet; had come to be resigned, she said (and I believe her); and had a mind at peace. ' In a word, you are happy here ?' said one of my companions. She struggled—she did struggle very hard—to answer, Yes: but raising her eyes, and meeting that glimpse of freedom over-head, she burst into tears, and said, ' She tried to be ; she uttered no complaint; but it was natural that she should sometimes long to go out of that one cell : she could not help *that*,' she sobbed, poor thing !

"I went from cell to cell that day ; and every face I saw, or word I heard, or incident I noted, is present to my mind in all its painfulness. But let me pass them by, for one, more pleasant, glance of a prison on the same plan which I afterwards saw at Pittsburgh.

" When I had gone over that, in the same manner, I asked the governor if he had any person in his charge who was shortly going out. He had one, he said, whose time was up next day; but he had only been a prisoner two years.

" Two years ! I looked back through two years in my own life —out of jail, prosperous, happy, surrounded by blessings, comforts, and good fortune—and thought how wide a gap it was, and how long those two years passed in solitary captivity would have

been. I have the face of this man, who was going to be released
next day, before me now. It is almost more memorable in its hap-
piness than the other faces in their misery. How easy and how
natural it was for him to say that the system was a good one; and
that the time went ' pretty quick—considering;' and that when a
man once felt he had offended the law, and must satisfy it, ' he
got along, somehow :' and so forth !

" ' What did he call you back to say to you, in that strange
flutter ?' I asked of my conductor, when he had locked the door
and joined me in the passage.

" ' Oh ! that he was afraid the soles of his boots were not fit for
walking, as they were a good deal worn when he came in ; and that
he would thank me very much to have them mended, ready.'

" Those boots had been taken off his feet, and put away with the
rest of his clothes, two years before !

" I took that opportunity of inquiring how they conducted themselves immediately before going out ; adding that I presumed they
trembled very much.

" ' Well, it's not so much a trembling,' was the answer—' though
they do quiver—as a complete derangement of the nervous system.
They can't sign their names to the book ; sometimes can't even
hold the pen ; look about 'em without appearing to know why, or
where they are ; and sometimes get up and sit down again twenty
times in a minute. This is when they're in the office, where they
are taken with the hood on, as they were brought in. When they
get outside the gate, they stop, and look first one way and then
the other : not knowing which to take. Sometimes they stagger
as if they were drunk, and sometimes are forced to lean against
the fence, they're so bad :—but they clear off in course of time.'

* * * * * * *

" On the haggard face of every man among these prisoners, the
same expression sat. I know not what to liken it to. It had
something of that strained attention which we see upon the faces
of the blind and deaf, mingled with a kind of horror, as though
they had all been secretly terrified.

" In every little chamber that I entered, and at every gate
through which I looked, I seemed to see the same appalling coun-

tenance. It lives in my memory, with the fascination of a remarkable picture. Parade before my eyes, a hundred men, with one among them newly released from this solitary suffering, and I would point him out.

"The faces of the women, as I have said, it humanizes and refines. Whether this be, because of their better nature, which is elicited in solitude, or because of their being gentler creatures, of greater patience and longer suffering, I do not know; but so it is. That the punishment is nevertheless, to my thinking, fully as cruel and as wrong in their case, as in that of the men, I need scarcely add.

" My firm conviction is, that independent of the mental anguish it occasions—an anguish so acute and so tremendous, that all imagination of it must fall far short of the reality—it wears the mind into a morbid state, which renders it unfit for the rough contact and busy action of the world. It is my fixed opinion, that those who have undergone this punishment, MUST pass into society again morally unhealthy and diseased. There are many instances on record, of men who have chosen, or have been condemned, to lives of perfect solitude, but I scarcely remember one, even among sages of strong and vigorous intellect, where its effect has not become apparent, in some disordered train of thought, or some gloomy hallucination. What monstrous phantoms, bred of despondency and doubt, and born and reared in solitude, have stalked upon the earth, making creation ugly, and darkening the face of Heaven ! ·

" Suicides are rare among these prisoners : are almost, indeed, unknown. But no argument in favor of the system, can reasonably be deduced from this circumstance, although it is very often urged. All men who have made diseases of the mind, their study, know perfectly well that such extreme depression and despair as will change the whole character, and beat down all its powers of elasticity and self-resistance, may be at work within a man, and yet stop short of self-destruction. This is a common case.

" That it makes the senses dull, and by degrees impairs the bodily faculties, I am quite sure. I remarked to those who were with me in this very establishment at Philadelphia, that the criminals who had been there long, were deaf. They, who were in the habit of seeing these men constantly, were perfectly amazed at the idea, which they regarded as groundless and fanciful. And yet the very first prisoner to whom they appealed—one of their own selection—confirmed my impression (which was unknown to him) instantly, and said, with a genuine air it was impossible to doubt, that he couldn't think how it happened, but he *was* growing very dull of hearing.

"That it is a singularly unequal punishment, and affects the worst man least, there is no doubt. In its superior efficiency as a means of reformation, compared with that other code of regulations which allows the prisoners to work in company without communicating together, I have not the smallest faith. All the instances of reformation that were mentioned to me, were of a kind that might have been—and I have no doubt whatever, in my own mind, would have been—equally well brought about by the Silent System. With regard to such men as the negro burglar and the English thief, even the most enthusiastic have scarcely any hope of their conversion.

"It seems to me that the objection that nothing wholesome or good has ever had its growth in such unnatural solitude, and that even a dog or any of the more intelligent among beasts, would pine, and mope, and rust away, beneath its influence, would be in itself a sufficient argument against this system. But when we recollect, in addition, how very cruel and severe it is, and that a solitary life is always liable to peculiar and distinct objections of a most deplorable nature, which have arisen here; and call to mind, moreover, that the choice is not between this system and a bad or ill-considered one, but between it and another which has worked well, and is, in its whole design and practice, excellent; there is surely more than sufficient reason for abandoning a mode of punishment attended by so little hope or promise, and fraught, beyond dispute, with such a host of evils."

PHILADELPHIA, Jan. 20, 1845.

To WILLIAM PETER, Esq.,

Her Britannic Majesty's Consul-General for the State of Pennsylvania.

MY DEAR SIR,—You informed me some time ago, that you were satisfied, from repeated visits to the Eastern Penitentiary, that Mr. *Charles Dickens's* account of that institution in his *American Notes,* was exceedingly erroneous. You will confer, I think, a benefit upon the cause of truth as well as philanthropy, if you will communicate to me for publication, the result of your inquiries as to his facts, and your views of the soundness or fallacy of his general conclusions.

Samuel R. Wood, the former warden of this prison, has lately returned from England. He tells me that the honest repute of this eminent establishment, has been injured there by the representations of *Mr. Dickens,* whose note as a writer of fictions, has secured for his crude performance a diffusive popularity. What from the extravagant fancies of this writer on the one hand, and the inflamed party zeal of the Boston Prison Society's Reports on the other, the benevolent public, both at home and abroad, are in danger of being greatly abused and misled.

On such a question, your testimony and judgment as a gentleman of profound and various research, as an unbiassed foreigner of long acquaintance with prison discipline, would be of signal worth. I venture therefore to appeal to you, from the distortions of one whose native temperament gives him, perhaps, even less claim to consideration as a judge, than his very hurried and superficial inspection of the prison, entitles him to respect as a witness.

I am, very truly,

Yours, &c.,

J. R. TYSON.

PHILADELPHIA, January 25, 1845.

To JOB R. TYSON, Esq.

MY DEAR SIR,

I have received your letter of the 20th, respecting the Eastern Penitentiary of this city; and, in compliance with your request, as well as in justice to that institution and the benevolent individuals who superintend and conduct it, hasten to give you the result of the investigations, which, in consequence of Mr. Dickens' statements, I considered it my duty to make on the subject. Though I had frequently visited the penitentiary and approved of it as a whole, it was not until after the appearance of his remarks, that my attention was called to its more particular cases and details. The result of these subsequent and minuter inquiries has only served to strengthen and confirm my earlier impressions. Better arranged buildings, more judicious regulations, or humaner treatment of prisoners,—in short, means better adapted and directed to their proposed end,—I have never seen in any institution for the punishment and reformation of criminals. In truth, I might add, that it is *superior* to any thing of the kind that I am acquainted with, either in the old world or the new.

I. "*The first man*," noticed by Mr. D——, had come into the penitentiary in February, 1839, and left it February, 1843, the remaining portion of his sentence having been remitted. During his imprisonment he had been allowed to correspond with his wife,— a most respectable woman,—who supported herself and children by needle-work, and whose letters to her husband, were full of kind and excellent advice. On quitting prison he received $51 for extra work, and now earns a comfortable livelihood by his labours as a journeyman printer. As far as I am able to learn, he is not worse for his imprisonment, either in body or mind,— nay, as to the latter, very much improved. He is in correspondence with the chaplain, and writes a very good letter.

II. *The "German"* (who has ornamented his cell, and laid out the few feet of ground behind with such ingenuity and neatness) came in May, 1840, and will leave in May, 1845. He had been convicted of two offences, for each of which he was condemned to two years and an half imprisonment. The sentence has been considered by some as too severe; but as for his being a "dejected, heart-broken, wretched creature;" as for his "forlorn affliction and distress of mind," I could discover no signs or symptoms of either. He was in as excellent health and spirits as mortal need be,—conversed freely about his situation, and expressed confident hopes, that he should, through the kindness and recommendations of the governor and others, be able to get into good employment as a paper-stainer, on the expiration of his term of imprisonment. He is an ingenious and clever fellow, but a great hypocrite, and evidently saw Mr. D.'s *weak* side—saw

> " Drops of compassion trembling on his eyelids,
> Ready to fall, as soon as he had told his
> Pitiful story."

III. " *The black burglar*"—came in April, 1837—went out April, 1842—came in again on the 13th of July following. Just as Mr. Dickens described him, " a very dexterous thief; notorious for his boldness, hardihood, and for the number of his previous convictions," &c. He had been convicted of stealing silver spoons, and seemed to glory in the crime, telling me that, though bred to the *iron* trade, he liked the *silver* trade much better—scorns to be thought a common thief, and calls himself a burglar by profession—has a mania for plunder that can never be cured. He is one of those who " laugh and grow fat" in spite of all punishment.

IV. *The man "allowed to keep rabbits,"* came in November, 1833, and went out in November, 1842, in good health and spirits. He now resides in Canada, and (according to letters received from him by his countrymen) is doing well.

V. *The " English Thief,"* looks in good health and is conducting himself well.

VI. *The " Poet"* came in July, 1840, and left in July, 1843. He had been discarded by his father some years before, for intemperate habits ; he received on quitting prison $30 for extra work, besides $50 for the copyright of his book. He is now in respectable business, reconciled to his father, and respectably

s

married. (His wife knew of his imprisonment.) He frequently visits the warden, and is, to all appearance, well in mind, body, and circumstances.

VII. *The "accomplished surgeon,"* came in July, 1840, and left in January, 1843, in good health—he is now employed in a large apothecary's establishment in South America, and conducting himself with propriety. He has written to the chaplain of the penitentiary, thanking him and the officers of the prison, for their kindness to him during his confinement.

VIII. *The "pretty coloured boy,"* came in November, 1841, and left in November, 1843. He was quite ignorant and uninstructed when he entered, but learned whilst in prison to read, write, and cipher; has now a good place as servant, in Mr. ——'s family, and behaves remarkably well.

IX. *The " Sailor,"* came in December, 1830, and left in January, 1842. He had been convicted of rape—he left with no appearance about him of "the helpless, crushed, and broken man," but in apparent health and spirits. His first request, on being liberated, was to have "a chew of tobacco." He is now in the employment of a farmer in the interior of the State, and said to be conducting himself well.

X. *The "three young women in adjoining cells,"* still continue in prison, but have nothing "very sad" in their looks, or in any way calculated to move "the sternest visiter to tears." They have been a kind of decoy-ducks for keepers of low brothels, and were convicted of a conspiracy to rob their prosecutor. They came into prison quite ignorant and untaught, but now read, write, cipher, and work remarkably well. One of them (she to whom Mr. Dickens more particularly refers) told me that their imprisonment had been "a very good thing" for them all, and that she did not know what would have become of them, had they not been sent there—that they had been very bad girls, and used to be drunk from morning to night—and indeed, "had no comfort or peace except when drunk." She hopes now that she shall be able to earn an honest livelihood. Her parents (who are respectable coloured people in another State, and from whom she ran away at fifteen) are now reconciled and have written to say that they will receive, and do what they can for her when she comes out of prison. She has become an excellent seamstress, and they are now all three out of prison, in good service, and said to be conducting themselves with propriety.

I could not perceive that any of " the criminals, who had been long there, were deaf," or even more " dull of hearing" than the inmates of other prisons.

I have only to add, that, though I have frequently visited the penitentiary and seen and conversed with many of its inmates, I cannot recollect having witnessed a single instance of the pains and wretchedness described by Mr. Dickens. It is not true, that the prisoner " never hears of wife or children ; home or friends; the life or death of any single creature"—that, with the exception of the prison officers, " he never looks upon a human countenance or hears a human voice," &c. On the contrary, he is allowed, under proper restrictions, to correspond with, and even in some cases, to see both wife and children. He sees also, from time to time, moral instructors, and other benevolent individuals who are in the habit of visiting the prison,* and is always at liberty to have the minister of his own church or sect with him, except after lock-up hours, or when engaged in the daily task of the establishment. It is not, properly speaking, *solitary* imprisonment that he undergoes, but merely *separation* from his fellows in crime.

I have heard Mr. Dickens accused of wilful misrepresentation. Of that I most fully absolve him. I do not think that he would be guilty—knowingly guilty—of a falsehood for any consideration. But all things are not given to all men ; and the very

[* Mr. Peter's reference includes the members of the Acting Committee of the Philadelphia Prison Society. How freely the intercourse of convicts with honest people may be encouraged without a violation of the "SEPARATE SYSTEM" may be seen in the fact that during the last year alone, members of the Acting Committee have made more than seven hundred and fifty visits to the Eastern Penitentiary, during which they held more than seven thousand interviews with the prisoners, averaging at least a quarter of an hour each. Of these visits, more than one hundred and fifty written reports were made to the Committee on the Penitentiary. During the same period a committee of educated women made nearly twelve hundred visits to the convicts in the female departments of the Penitentiary and the Moyamensing Prison.]

faculty which has enabled him so to excel in one species of composition, almost incapacitates him for some others. His prison-scenes are much of a kin to Sterne's. Still I believe that he never deceived another without having first deceived himself.*

I am, my dear sir,

Very truly, yours, &c.,

WILLIAM PETER.

* For some admirable remarks on Mr. D., and on his merits and defects as a writer, see the LEAGUE newspaper of December 21st, p. 204.

www.ingramcontent.com/pod-product-compliance
Lightning Source LLC
Chambersburg PA
CBHW030630270326
41927CB00007B/1382